The Role of Network Security and 5G Communication in Smart Cities and Industrial Transformation

Edited by

Devasis Pradhan
Department of Electronics & Communication Engineering
Acharya Institute of Technology, Bangalore
Karnataka, India

Mangesh M. Ghonge
Department of Computer Engineering
Sandip Institute of Technology and Research Center
Nashik, India

Nitin S. Goje
Department of Management & Technology
Webster University, Tashkent, Uzbekistan

Alessandro Bruno
Department of Computing and Informatics
Bournemouth University, United Kingdom

&

Rajeswari
Department of Electronics & Communication Engineering
Acharya Institute of Technology, Bangalore
Karnataka, India

The Role of Network Security and 5G Communication in Smart Cities and Industrial Transformation

Editors: Devasis Pradhan, Mangesh M. Ghonge, Nitin S. Goje, Alessandro Bruno and Rajeswari

ISBN (Online): 978-981-5305-87-6

ISBN (Print): 978-981-5305-88-3

ISBN (Paperback): 978-981-5305-89-0

need for a court order if at any point you breach any terms of this License Agreement. In no event will any delay or failure by Bentham Science Publishers in enforcing your compliance with this License Agreement constitute a waiver of any of its rights.

3. You acknowledge that you have read this License Agreement, and agree to be bound by its terms and conditions. To the extent that any other terms and conditions presented on any website of Bentham Science Publishers conflict with, or are inconsistent with, the terms and conditions set out in this License Agreement, you acknowledge that the terms and conditions set out in this License Agreement shall prevail.

Bentham Science Publishers Pte. Ltd.
80 Robinson Road #02-00
Singapore 068898
Singapore
Email: subscriptions@benthamscience.net

BENTHAM SCIENCE

CONTENTS

PREFACE

In the ever-evolving landscape of technology, the intersection of network security and 5G communication stands at the forefront of driving significant transformations in our societies. As we navigate the era of smart cities and the industrial revolution, the integration of these two crucial elements plays a pivotal role in shaping the future of connectivity, efficiency, and innovation. "The Role of Network Security and 5G Communication in Smart Cities and Industrial Transformation" delves into the intricate relationship between the security of our digital infrastructure and the revolutionary capabilities of fifth-generation (5G) communication technologies. This preface serves as a gateway to understanding the complex dynamics at play and exploring the challenges and opportunities that arise in the context of smart cities and industrial metamorphosis.

Smart cities are emerging as hubs of interconnected technologies, where data-driven decision-making, automation, and connectivity are redefining urban living. The synergy between 5G communication and network security becomes paramount in ensuring the seamless operation of diverse applications, ranging from smart grids and intelligent transportation systems to healthcare and public safety. This preface sets the stage for an in-depth exploration of how robust network security becomes the bedrock upon which the promises of 5G in smart cities can be fully realized. Simultaneously, as industries undergo a profound transformation with the advent of Industry 4.0, characterized by the fusion of digital technologies, the role of network security becomes even more critical. The deployment of 5G communication networks in industrial settings promises unprecedented gains in efficiency, productivity, and flexibility. However, this also introduces new vulnerabilities that require meticulous attention to safeguard critical infrastructure, intellectual property, and sensitive data. This preface aims to articulate the delicate balance required to harness the potential of 5G in industrial applications while fortifying digital defenses against evolving cyber threats.

As we embark on this exploration, the preface provides a roadmap for readers, outlining the key themes, objectives, and significance of the ensuing chapters. It encourages a holistic understanding of the intricate interplay between network security and 5G communication in the context of smart cities and industrial transformation. Together, these elements converge to shape a future where connectivity is not only ubiquitous but also secure, empowering societies to embrace the transformative potential of technology while safeguarding against the challenges that accompany progress.

Devasis Pradhan
Department of Electronics & Communication Engineering
Acharya Institute of Technology, Bangalore
Karnataka, India

Mangesh M. Ghonge
Department of Computer Engineering
Sandip Institute of Technology and Research Center
Nashik, India

Nitin S. Goje
Department of Management & Technology
Webster University, Tashkent, Uzbekistan

Alessandro Bruno
Department of Computing and Informatics
Bournemouth University, United Kingdom

&

Rajeswari
Department of Electronics & Communication Engineering
Acharya Institute of Technology, Bangalore
Karnataka, India

INTRODUCTION

In the dynamic landscape of technology and urban development, the synergy between network security and 5G communication stands as a linchpin for transformative change. This book, titled "The Role of Network Security and 5G Communication in Smart Cities and Industrial Transformation," embarks on a comprehensive exploration of the intricate relationship between these two pillars, unraveling the profound impact they wield in shaping the future of our cities and industries. As we navigate the era of unprecedented technological advancements, the concept of smart cities emerges as a beacon of innovation. Smart cities represent a holistic integration of digital technologies, data-driven insights, and intelligent infrastructure, promising to revolutionize the way we live, work, and interact with our urban environments. At the same time, industries are experiencing a paradigm shift with the advent of Industry 4.0, where automation, connectivity, and data analytics converge to redefine traditional manufacturing processes. Both of these revolutions are underpinned by the transformative potential of 5G communication networks.

This book serves as a guide to unravel the symbiotic relationship between network security and 5G communication in the context of smart cities and industrial transformation. The deployment of 5G networks brings forth unparalleled connectivity, enabling faster data speeds, lower latency, and the ability to connect a multitude of devices simultaneously. However, this newfound connectivity also presents novel challenges in terms of cybersecurity, privacy, and the integrity of critical systems. The pages of this book unfold the critical role of network security in mitigating these challenges, ensuring that the promises of 5G can be harnessed securely and responsibly.

The primary aim of this book is to provide a nuanced understanding of the key themes and challenges within the realm of network security and 5G communication. The objectives include:

Smart Cities and Urban Dynamics: Delving into the impact of 5G and network security on the development of smart cities, exploring the challenges and opportunities that arise in creating intelligent, responsive urban environments.

Industrial Evolution with 5G: Investigating the transformative potential of 5G in industrial settings, uncovering how network security becomes paramount in ensuring the reliability and security of interconnected industrial systems.

Cybersecurity Challenges: Identifying and dissecting the cybersecurity challenges inherent in the 5G landscape, offering insights into proactive measures and best practices to safeguard against evolving threats.

Ethical and Responsible Implementation: Advocating for the ethical and responsible implementation of 5G technologies, considering the implications for privacy, data protection, and the overall well-being of individuals and communities.

The book endeavors to provide readers with a comprehensive framework for understanding the intertwined dynamics of network security and 5G communication. By doing so, it aims to equip professionals, researchers, policymakers, and enthusiasts alike with the knowledge necessary to navigate the complexities of our evolving digital landscape while fostering a secure and resilient future for smart cities and industrial realms alike.

List of Contributors

Alessandro Bruno	Department of Computing and Informatics, Bournemouth University, United Kingdom
Akash Kotagi	Department Electronics and Communication Engineering, Rashtreeya Vidyalaya College of Engineering, Bengaluru, India
Amit Kumar Sahoo	Lead 1 Workforce Management, UST Global, Bangalore, India
Anupam Mukherjee	Department of Health and Family Welfare, West Bengal Homoeopathic Health Service, Government of West Bengal, India
Anita Sardar Patil	Bharati Vidyapeeth (Deemed to be University) Homoeopathic Medical College, , Pune, India
B. Sahana	Department of Electronics and Communication, R. V. College of Engineering Bangalore-560059, India
B. Sadhana	Department of Electronics and Communication, Canara College of Engineering, Mangalore, India
Bhanudas Suresh Panchbhai	Department of Computer Science, R.C. Patel Arts, Commerce and Science College, Shirpur, Maharashtra, India
C.S. Meghana	Department of Electronics and Communication, R. V. College of Engineering Bangalore-560059, India
Devasis Pradhan	Department of Electronics & Communication Engineering, Acharya Institute of Technology, Bangalore, Karnataka, India
Dhanush Prabhakar	Department of Electronics and Communication, R. V. College of Engineering Bangalore-560059, India
Kishan Gupta	Department of Electronics & Communication Engineering, C. V. Raman Global University, Bhubaneswar, Odisha-752054, India
Prasanna Kumar Sahu	Department of Electrical Engineering, National Institute of Technology, Rourkela-769008, Odisha, India
Prabhakar Rath	Department of Electronics & Communication Engineering, C. V. Raman Global University, Bhubaneswar, Odisha-752054, India
Pushpendra Pal Singh	G.L. Bajaj Institute of Management, Greater Noida, India
P. Kalyan Ram	Department Electronics and Communication Engineering, Rashtreeya Vidyalaya College of Engineering, Bengaluru, India
Rakesh Kumar Dixit	G.L. Bajaj Institute of Management, Greater Noida, India
Smita Rani Parija	Department of Electronics & Communication Engineering, C. V. Raman Global University, Bhubaneswar, Odisha-752054, India
Sindhu Rajendran	Department Electronics and Communication Engineering, Rashtreeya Vidyalaya College of Engineering, Bengaluru, India
Sachin Kadam	Institute of Management and Entrepreneurship Development, Bharati Vidyapeeth (Deemed to be University), Pune, India
Tarique Akhtar	Data Science Agility, Dubai

Umesh Ghate	Bharati Vidyapeeth (Deemed to be University), College of Ayurved, Pune, India
Varsha Umesh Ghate	Bharati Vidyapeeth (Deemed to be University) Homoeopathic Medical College, , Pune, India
Varsha Makarand Pathak	Department of Computer Applications, KCES'S Institute of Management and Research, Maharashtra, India

CHAPTER 1

Sustainability in Smart Cities: A 5G Green Network Approach

Devasis Pradhan[1,*], Prasanna Kumar Sahu[2] and Alessandro Bruno[3]

[1] Department of Electronics & Communication Engineering, Acharya Institute of Technology, Bangalore, Karnataka, India

[2] Department of Electrical Engineering, National Institute of Technology, Rourkela-769008, Odisha, India

[3] Department of Computing and Informatics, Bournemouth University, United Kingdom

Abstract: The rapid urbanization and technological advancements of the 21st century have propelled the evolution of smart cities, aiming to enhance efficiency, connectivity, and overall quality of life. As cities strive to address environmental challenges, this research investigates the integration of a 5G Green Network as a pivotal component of smart city sustainability. The study explores the intersection of 5G technology and environmentally conscious practices, aiming to understand their collective impact on urban development. The literature review underscores the current landscape of smart cities, sustainability, and the emergent role of 5G networks. Highlighting gaps in existing research, the paper establishes the need for an in-depth examination of the potential environmental benefits and challenges associated with deploying 5G technology in smart city infrastructures. A conceptual framework is proposed, delineating the key components of a 5G Green Network and its seamless integration into smart city infrastructure. The methodology section outlines research design, data collection methods, and analytical tools employed to assess the sustainability implications of 5G technology. The paper examines the various facets of smart city infrastructure and elaborates on how 5G Green Networks can positively impact energy efficiency, reduce carbon emissions, and enhance overall sustainability. Drawing on case studies and examples, the research presents successful instances of cities implementing 5G Green Networks and analyzes the lessons learned. This research aims to provide valuable insights for policymakers, urban planners, and technologists alike, fostering a deeper understanding of the potential of 5G Green Networks in advancing the sustainability agenda within the context of smart cities.

Keywords: IoT, Green technology, Green communication, Smart cities, 5G Network.

---

* **Corresponding author Devasis Pradhan:** Department of Electronics & Communication Engineering, Acharya Institute of Technology, Bangalore, Karnataka, India; E-mail: devasispradhan@acharya.ac.in

Devasis Pradhan, Mangesh M. Ghonge, Nitin S. Goje, Alessandro Bruno and Rajeswari (Eds.)

INTRODUCTION

In the midst of an era characterized by rapid urbanization and burgeoning technological innovation, the concept of smart cities has emerged as a transformative paradigm for urban development. Smart cities leverage cutting-edge technologies to enhance efficiency, connectivity, and overall livability, aiming to create urban ecosystems that respond intelligently to the needs of their inhabitants. This evolution towards smart urbanization, however, is not without its challenges, and one of the paramount concerns is the imperative of sustainability. Sustainability in the context of smart cities goes beyond mere ecological considerations; it encompasses a holistic approach that integrates environmental responsibility, economic viability, and social inclusivity. The need for sustainable urban development has become increasingly urgent in the face of climate change, resource constraints, and the burgeoning global population. As we strive to construct cities that endure, fostering a harmonious coexistence between humanity and the environment becomes paramount.

This research paper embarks on an exploration of the interplay between sustainability and smart cities, with a specific focus on the transformative potential of 5G green networks. The research problem at the heart of this inquiry lies in understanding how the deployment of 5G technology, coupled with environmentally conscious practices, can contribute to the sustainable development of smart cities. As traditional telecommunication networks pave the way for 5G, a convergence of connectivity and environmental responsibility presents itself as an opportunity to reshape the urban landscape. The importance of incorporating 5G Green Networks in smart city infrastructure cannot be overstated. Beyond the anticipated advancements in communication speeds and data capacity, 5G networks hold the promise of reduced energy consumption and a diminished carbon footprint. This paradigm shift from conventional networks to green, energy-efficient alternatives underscores the potential for 5G technology to be a catalyst for environmental sustainability within the urban context.

Literature Review

The literature on smart cities elucidates the multifaceted nature of urban development, emphasizing the integration of information and communication technologies (ICTs) to enhance the efficiency and quality of urban services. Scholars such as Caragliu, A., Del Bo, C., & Nijkamp, P [1]. and Hollands [2] have extensively examined the concept, highlighting the potential for smart cities to improve resource allocation, environmental sustainability, and overall urban governance. Sustainability in urban development has been a recurring theme in the literature, with researchers emphasizing the need to balance economic growth with environmental responsibility and social equity. Works by Beatley [3] and Newman and Jennings [3] underscore the importance of creating cities that are resilient, resource-efficient, and inclusive, considering the ecological impact of urbanization. Recent studies exploring the integration of 5G technology into urban environments have primarily focused on the anticipated advancements in communication speeds and data capacity. Notable contributions by Zhang *et al.* [4] and Misra *et al.* [5] provide insights into the technical aspects of 5G deployment, emphasizing its potential to revolutionize connectivity and enable new applications across various sectors. Despite the wealth of literature on smart cities, sustainability, and 5G technology, a critical analysis reveals discernible gaps that necessitate further exploration. First and foremost, the intersection of sustainability, smart cities, and 5G networks remains underexplored. Few studies have comprehensively addressed the potential environmental impact of 5G technology in the broader context of urban sustainability. This research is poised to bridge these gaps by providing a holistic examination of the integration of 5G Green Networks in smart cities. By weaving together insights from the realms of smart cities, sustainability, and 5G technology, this study aims to contribute a nuanced understanding of the potential environmental benefits and challenges associated with 5G deployment in urban contexts.

Conceptual Framework

The conceptual framework provides a structured basis for the subsequent analysis and discussion of the research findings, offering a visual representation of the interdependencies that define the integration of 5G Green Networks in the broader context of smart city sustainability shown in Fig. (**1**) [6].

The fundamental key terms associated with the framework is as follows:

- **Smart Cities:** Smart cities leverage information and communication technologies (ICTs) to enhance urban infrastructure, services, and the overall quality of life for residents. This includes the integration of data-driven solutions for efficient governance, sustainable resource management, and improved connectivity.
- **Sustainability:** Sustainability in the context of urban development refers to the balanced integration of economic, environmental, and social considerations. It involves creating cities that meet the needs of the present without compromising the ability of future generations to meet their own needs [7].
- **5G Technology:** 5G technology represents the fifth generation of mobile networks, characterized by significantly faster data transfer speeds, lower latency, and increased capacity compared to previous generations. It forms the

backbone for advanced applications such as the Internet of Things (IoT), artificial intelligence (AI), and augmented reality (AR) [8, 9].

- **Green Networks:** Green networks, in the context of this conceptual framework, refer to telecommunication networks designed with a focus on environmental sustainability. This includes reducing energy consumption, minimizing carbon emissions, and adopting eco-friendly practices in the deployment and maintenance of network infrastructure.

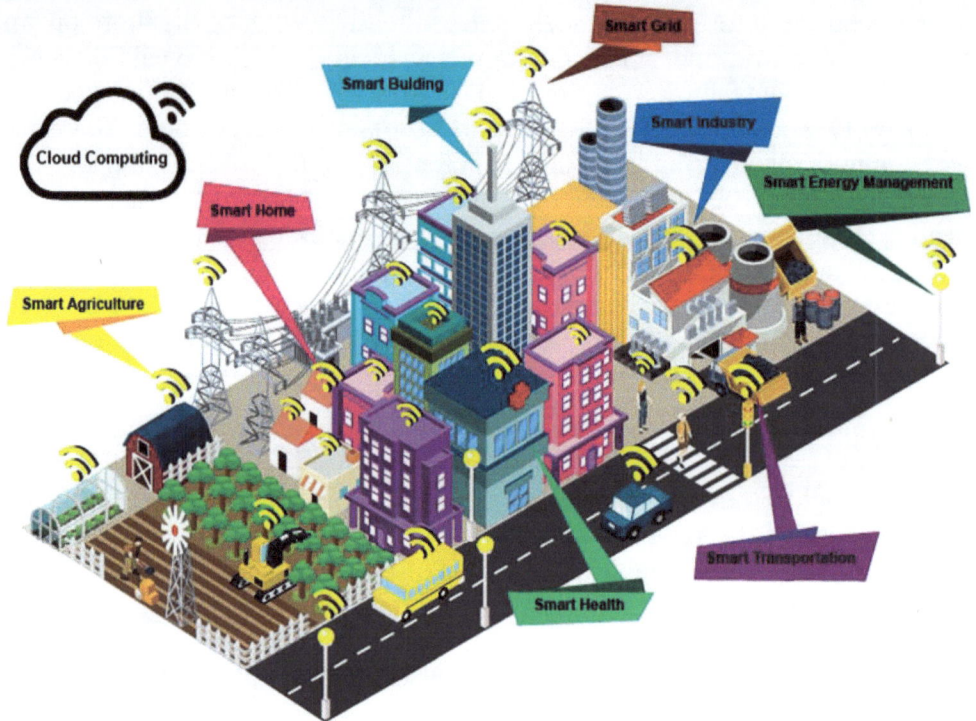

Fig. (1). 5G enabled smart city.

The conceptual framework for the integration of 5G Green Networks in smart cities is structured around three interconnected pillars: Urban infrastructure encompasses various elements such as transportation systems, energy grids, water management, waste disposal, and public services. 5G technology serves as the underlying connectivity layer, facilitating real-time communication and data exchange between different components of the smart city infrastructure [10]. The deployment of 5G Green Networks involves the implementation of energy-efficient network infrastructure, including base stations, antennas, and data centers. To minimize the environmental impact, 5G Green Networks can integrate renewable energy sources such as solar and wind power to meet their energy demands. The conceptual framework evaluates the environmental benefits of 5G Green Networks, including reduced energy consumption, lower carbon emissions, and overall ecological sustainability [11]. Considerations of the economic feasibility and long-term financial viability associated with the deployment of 5G Green Networks in smart city infrastructures. Analysis of the social implications, including increased accessibility, inclusivity, and potential improvements in the quality of life for urban residents. The framework emphasizes the symbiotic relationship between smart city infrastructure and 5G Green Networks. The integration of energy-efficient, environmentally conscious network technologies enhances the overall sustainability of smart city initiatives [12]. Furthermore, the environmental, economic, and social impact assessment provides a comprehensive understanding of the holistic benefits and challenges associated with the deployment of 5G Green Networks in smart cities.

SMART CITY INFRASTRUCTURE AND 5G GREEN NETWORKS

Components of Smart City

Implementing 5G technology in smart city infrastructure enhances real-time communication and data exchange between autonomous vehicles, traffic management systems, and smart transportation solutions. This integration allows for dynamic traffic management, optimized routing, and improved overall efficiency. 5G networks facilitate the deployment of smart grids and advanced energy management systems. These systems leverage the high data speeds and low latency of 5G to monitor and control energy distribution, optimize grid performance, and enable more efficient consumption patterns. 5G enhances the capabilities of surveillance and security systems through high-definition video streaming, real-time analytics, and quick response mechanisms [13, 14]. This integration improves situational awareness, aids in crime prevention, and enhances emergency response. 5G supports the development of telemedicine and remote patient monitoring, enabling healthcare systems to provide more accessible and efficient services. High-speed, low-latency connections enable real-time communication between healthcare professionals and patients, regardless of geographical distances. 5G technology allows for the deployment of smart waste management systems. Bin sensors, waste collection vehicles, and disposal facilities can be connected through 5G networks, optimizing waste collection routes and reducing operational costs. Smart water management systems benefit from 5G connectivity by enabling real-time monitoring of water quality, leak detection, and efficient distribution. This integration contributes to more sustainable water usage and reduces losses in the distribution network. 5G serves as the backbone of communication and connectivity in smart cities, supporting a myriad of IoT devices, sensors, and smart applications [15]. This connectivity enables seamless communication between various components of the smart city infrastructure. Fig. (**2**) discuss about the components of Smart Cities Infrastructure.

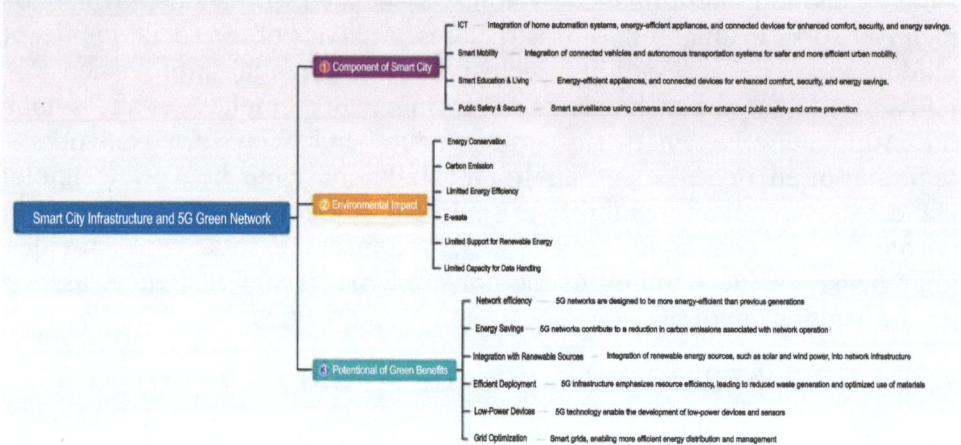

Fig. (2). Component of smart city infrastructure.

Environmental Impact of Traditional Networks

Conventional networks, including 3G and 4G, often require significant energy consumption for data transmission and maintenance. The energy-intensive nature of these networks contributes to a substantial carbon footprint [16]. Earlier generations of networks may face limitations in terms of data capacity and efficiency, resulting in potential bottlenecks, slower data transfer speeds, and less optimal performance for smart city applications. Higher latency in traditional networks may impede the real-time responsiveness required for applications like autonomous vehicles, healthcare systems, and critical infrastructure management [17, 18].

Potential Green Benefits of 5G Technology

5G technology, especially when designed with energy efficiency in mind, can significantly reduce energy consumption compared to traditional networks. Advanced features like network slicing and dynamic energy management contribute to improved efficiency. By leveraging renewable energy sources and optimizing energy use, 5G Green Networks have the potential to reduce carbon emissions associated with network operations, making them more environmentally friendly [19, 20]. 5G Green Networks: The higher data capacity of 5G networks allows for more efficient data transfer, reducing the need for multiple data transmissions and resulting in a more streamlined and eco-friendly network. The low latency and high-speed capabilities of 5G contribute to improved efficiency and responsiveness in smart city applications. This enhanced connectivity enables quicker decision-making and more effective management of urban systems [21]. Optimized network design and resource utilization contribute to sustainability by minimizing unnecessary resource consumption during the manufacturing, deployment, and operational phases of the network. The extensive support for IoT devices in 5G networks allows for the proliferation of smart devices and sensors, enabling more precise and data-driven management of resources in a smart city. We explore the components of smart city infrastructure and how 5G technology can be integrated into these systems [22].

CASE STUDIES

In January 2022, specific case studies or examples of cities that have fully implemented 5G Green Networks in their smart city initiatives may be limited, as the widespread deployment of 5G technology was still in progress. Table **1** gives examples of cities that were actively exploring or piloting green initiatives within their 5G deployments.

Table 1. Case studies on 5G deployment.

S. No.	Region	Initiative	Sustainable Outcome	Remarks
1	Barcelona, Spain	Barcelona has been at the forefront of smart city development, and it has explored the integration of 5G technology to enhance sustainability. The city has experimented with 5G-powered applications in transportation, waste management, and energy efficiency.	The potential outcomes include optimized traffic flow, efficient waste collection, and improved energy management. These initiatives aim to contribute to reduced carbon emissions and increased overall environmental sustainability.	Barcelona's efforts emphasize the importance of multi-stakeholder collaboration, including partnerships between the public sector, private companies, and research institutions. Effective engagement with citizens and addressing privacy concerns are also crucial aspects of successful implementation.
2	Singapore	Singapore has been actively exploring the use of 5G technology for its Smart Nation initiative. While the deployment of 5G was ongoing, the city-state aimed to integrate green and sustainable practices into its smart city infrastructure.	Anticipated outcomes include improved energy efficiency in buildings, enhanced traffic management through smart transportation systems, and the implementation of smart grids for optimized energy distribution.	Singapore's approach highlights the importance of long-term planning and a commitment to sustainability from the initial stages of smart city development. The city emphasizes the need for a regulatory framework that supports innovation while ensuring environmental responsibility.

(Table 1) cont.....

S. No.	Region	Initiative	Sustainable Outcome	Remarks
3	Stockholm, Sweden	Stockholm has been exploring the potential of 5G technology in the context of its smart city initiatives. The city has been considering how 5G can contribute to sustainability goals, including energy efficiency and reduced environmental impact.	The expected outcomes include the use of 5G for smart grids, intelligent transportation systems, and efficient resource management. These applications aim to enhance sustainability and resilience in urban living.	Stockholm's case underscores the importance of continuous innovation and adaptation. As technology evolves, cities need to stay agile and be willing to adjust their strategies to meet changing sustainability goals.
4	Seoul, South Korea	South Korea, including the city of Seoul, has been a pioneer in 5G technology deployment. Seoul has been exploring the integration of 5G in various aspects of urban life, including transportation, healthcare, and public services.	While specific outcomes related to green initiatives are evolving, the potential benefits include energy-efficient transportation systems, smart energy grids, and optimized urban planning for reduced environmental impact.	Seoul's experience highlights the importance of robust infrastructure and strategic planning. Implementing 5G green initiatives requires a comprehensive understanding of the city's unique challenges and a willingness to invest in technological solutions that align with sustainability goals.

Successful implementation requires collaboration between government bodies, private sector partners, and research institutions. Engaging citizens and addressing their concerns is crucial for the success of any smart city initiative [23]. A supportive regulatory environment that encourages innovation while ensuring environmental responsibility is vital. Sustainable smart city initiatives require long-term planning and a commitment to ongoing innovation and adaptation.

CHALLENGES AND SOLUTIONS

The challenges focused on a multi-faceted and collaborative approach, involving governments, industry players, communities, and regulatory bodies. Continuous monitoring, flexibility in adapting strategies, and a commitment to sustainability are essential for the successful implementation of 5G Green Networks in smart cities. Table **2** discusses various challenges and solutions to it.

BENEFITS OF 5G GREEN NETWORKS

The incorporation of 5G Green Networks in smart cities brings about a range of benefits across environmental, economic, and social dimensions shown in Fig. (**3**). From energy efficiency and reduced emissions to economic growth and improved quality of life, these networks play a pivotal role in creating sustainable and technologically advanced urban environments [24, 25].

Table 2. Challenges associated with implementing 5g green networks in smart cities.

S. No.	Parameter	Challenges	Solution
1	High initial cost	The deployment of 5G Green Networks involves significant upfront costs for infrastructure upgrades, energy-efficient equipment, and integration with existing systems.	Governments and private entities can explore public-private partnerships to share the financial burden. Incentives, subsidies, or grants may be provided to encourage the adoption of green technologies.
2	Energy consumption & efficiency	The increased data processing capacity and higher density of connected devices in 5G networks may lead to higher energy consumption.	• Implement energy-efficient technologies, such as dynamic power management, and invest in renewable energy sources to power 5G infrastructure. • Continuous monitoring and optimization can ensure efficient energy use.
3	Integration with system	Integrating 5G Green Networks with existing legacy systems and infrastructure can be complex, leading to compatibility issues.	Develop clear transition plans and strategies for seamless integration. Invest in interoperable technologies and conduct thorough compatibility testing to minimize disruptions during the implementation process.
4	Security and privacy	The increased connectivity in smart cities raises concerns about data security and privacy.	• Implement robust security measures, encryption protocols, and secure network architectures. • Establish clear data privacy regulations and educate the public about the security measures in place to build trust.
5	Public awareness	Public resistance due to concerns about health effects, privacy, or perceived risks may hinder the adoption of 5G Green Networks.	• Conduct extensive public awareness campaigns to address concerns, provide transparent information about the technology's benefits, and involve communities in decision-making processes. • Address health concerns by adhering to safety guidelines and communicating the scientific consensus.

(Table 2) cont.....

S. No.	Parameter	Challenges	Solution
6	Lack of skilled workforce	The deployment of 5G Green Networks requires a skilled workforce, and there may be a shortage of professionals with the necessary expertise.	• Invest in training programs, educational initiatives, and partnerships with academic institutions to build a skilled workforce. • Collaborate with industry associations to create certification programs and promote knowledge exchange.
7	Environmental infrastructure deployment	The physical deployment of 5G infrastructure, including the manufacturing and transportation of equipment, can have environmental implications.	• Prioritize sustainable practices in infrastructure deployment, such as using eco-friendly materials, minimizing waste, and optimizing transportation routes. • Implement circular economy principles to maximize the lifespan and recyclability of network components.
8	Global supply chain	Global supply chain disruptions, as observed in various industries, can impact the timely availability of critical components for 5G Green Networks.	• Diversify supply chains, source components locally where possible, and establish contingency plans to mitigate the risks associated with supply chain disruptions. • Foster collaboration between governments and industry to ensure a resilient supply chain.
9	Regulatory hurdles	Regulatory frameworks may not be adapted to address the specific considerations of 5G Green Networks, leading to delays and uncertainties.	• Engage with regulatory bodies early in the planning stages, advocating for policies that support green initiatives. • Foster collaboration between governments, industry stakeholders, and regulatory bodies to develop guidelines that encourage sustainability.
10	Community inclusivity and accessibility	Ensuring that the benefits of 5G Green Networks reach all segments of the population can be challenging, leading to potential disparities.	• Develop inclusive policies and initiatives to address the digital divide. • Provide affordable access, especially in underserved areas, and actively engage with communities to understand and address their specific needs.

Environmental Benefit

5G Green Networks are designed to be more energy-efficient than their predecessors, optimizing power usage and reducing overall energy consumption. Advanced technologies, such as network slicing and dynamic resource allocation, contribute to improved efficiency. The deployment of 5G Green Networks can be coupled with the integration of renewable energy sources, such as solar and wind power [25]. This reduces dependence on non-renewable energy and lowers the carbon footprint of network operations.By minimizing energy consumption and integrating renewable energy, 5G Green Networks contribute to lower carbon emissions. The overall environmental impact is reduced, aligning with global efforts to combat climate change.

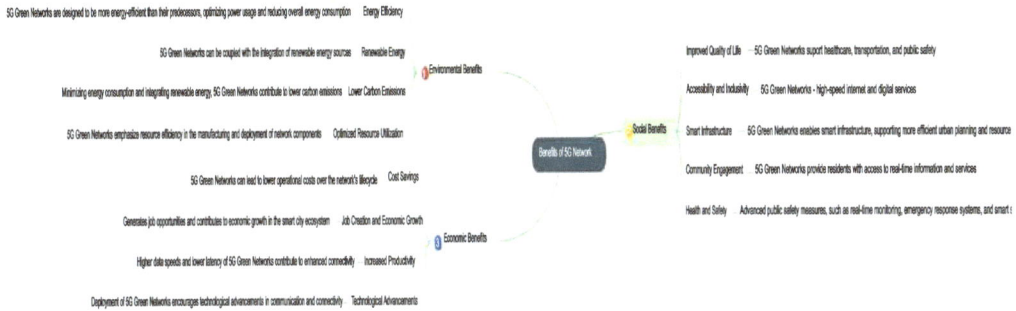

Fig. (3). Benefits of 5G green network.

Efficient Use of Materials

5G Green Networks emphasize resource efficiency in the manufacturing and deployment of network components. This leads to minimized waste generation and a more sustainable approach to resource utilization.

Economics Benefits

Through improved energy efficiency and optimized resource utilization, 5G Green Networks can lead to lower operational costs over the network's lifecycle. This is especially relevant in the long term, contributing to cost savings for both operators and end-users. The higher data speeds and lower latency of 5G Green Networks contribute to enhanced connectivity, enabling faster and more reliable communication. This, in turn, boosts productivity across various sectors, from healthcare to manufacturing. The deployment of 5G Green Networks encourages technological advancements in communication and connectivity. This stimulates innovation, creating opportunities for businesses and fostering a culture of continuous improvement.

Social Benefits

5G Green Networks support the delivery of improved services in areas such as healthcare, transportation, and public safety. Real-time communication and data exchange contribute to more responsive and efficient city services, enhancing the overall quality of life for residents.5G Green Networks facilitate broader connectivity, ensuring that more people have access to high-speed internet and digital services. This helps bridge the digital divide and promotes inclusivity, regardless of geographic location.

Efficient Urban Planning

The deployment of 5G Green Networks enables smart infrastructure, supporting more efficient urban planning and resource management. This, in turn, contributes to a more sustainable and livable urban environment.5G Green Networks provide residents with access to real-time information and services, empowering them to make informed decisions about their daily lives. This level of connectivity fosters community engagement and participation in city initiatives.

Enhanced Public Safety

Improved connectivity supports advanced public safety measures, such as real-time monitoring, emergency response systems, and smart surveillance. This contributes to enhanced safety and security for residents.

POLICY IMPLICATIONS

Implementing this comprehensive policy framework requires close collaboration between government agencies, private sector stakeholders, local communities, and research institutions. The policies should be adaptive, considering evolving technologies and sustainability goals while fostering an environment that encourages innovation and the responsible deployment of 5G Green Networks in smart cities.

Regulatory Guidelines

- **Emission standards:** Establish clear emission standards for 5G infrastructure to ensure that the deployment and operation of networks align with environmental goals. These standards may include limits on energy consumption and emissions associated with network components.
- **Compliance requirements:** Develop regulatory frameworks that mandate compliance with green standards for 5G networks. This may include periodic reporting on energy efficiency, emissions, and adherence to sustainability practices.

Incentive Programs

- **Financial incentives:** Introduce financial incentives, such as tax credits or subsidies, to encourage network operators and businesses to invest in 5G Green Networks. These incentives can offset the initial costs of infrastructure upgrades and deployment.

- **Performance-based incentives:** Tie incentives to specific performance metrics related to energy efficiency, emissions reduction, and the use of renewable energy sources. Reward operators for achieving and surpassing sustainability targets.

Public-Private Partnerships

- **Collaborative planning:** Foster collaboration between government bodies, private sector entities, and research institutions to develop comprehensive plans for the deployment of 5G Green Networks. Public-private partnerships can leverage the strengths of each sector to ensure successful implementation.
- **Joint investment initiatives:** Establish joint investment initiatives where public and private entities share the financial burden of deploying 5G infrastructure. This collaborative approach ensures a more equitable distribution of costs and benefits.

Research and Development Support

- **Funding for innovation:** Allocate funds for research and development initiatives focused on advancing green technologies within the 5G ecosystem. Government support can stimulate innovation in areas such as energy-efficient network components, sustainable materials, and renewable energy integration.
- **Technology testbeds:** Create testbeds and pilot projects that allow for the testing and validation of new green technologies within smart city environments. These initiatives can help identify best practices and potential challenges before full-scale deployment.

Sustainability Reporting Requirements

- **Transparency measures:** Implement requirements for network operators to regularly report on the environmental sustainability of their 5G networks. These reports can include metrics such as energy consumption, carbon emissions, and the use of renewable energy sources.
- **Public accessibility:** Ensure that sustainability reports are publicly accessible to promote transparency. This allows stakeholders, including citizens, to assess the environmental impact of 5G Green Networks and hold operators accountable.

Standardization and certification

- **Environmental standards:** Work with industry stakeholders to establish environmental standards for 5G network equipment and infrastructure. These standards can guide manufacturers in producing environmentally friendly components.

- **Certification programs:** Introduce certification programs that verify the compliance of 5G network equipment with established environmental standards. Certification can serve as a market incentive, signaling to operators and consumers that the equipment meets green criteria.

Community Engagement and Education

- **Public awareness campaigns:** Conduct public awareness campaigns to educate citizens about the benefits of 5G Green Networks and address any concerns related to environmental impact or health. Informed citizens are more likely to support and participate in sustainable smart city initiatives.
- **Community consultations:** Involve communities in the decision-making process related to the deployment of 5G Green Networks. Seek input on the placement of infrastructure, address community-specific needs, and ensure that the benefits of 5G are equitably distributed.

Digital Inclusion Policies

- **Access and affordability:** Develop policies that promote digital inclusion, ensuring that the benefits of 5G Green Networks are accessible to all socioeconomic groups. This may involve subsidies for low-income households, community access points, and initiatives to bridge the digital divide.
- **Inclusive service planning:** Work with operators to develop inclusive service plans that consider the diverse needs of the population. This may include customized packages for different demographics and a focus on providing essential services to underserved areas.

Flexible Zoning and Permitting

- **Streamlined approvals:** Establish streamlined zoning and permitting processes for the deployment of 5G infrastructure. This reduces bureaucratic hurdles and accelerates the implementation of green networks.
- **Flexible land use policies:** Allow for flexible land use policies that accommodate the installation of renewable energy sources, such as solar panels on network infrastructure, and support sustainable urban planning.

Cybersecurity and Privacy Regulations

- **Security standards:** Enforce cybersecurity standards for 5G networks to protect against potential threats. Governments should work with industry experts to establish robust security protocols.
- **Privacy protection:** Implement regulations that safeguard user privacy in the context of 5G networks. Ensure that data collection and processing adhere to strict privacy standards, promoting trust among users.

CONCLUSION

The integration of 5G Green Networks in smart cities represents a transformative leap toward a more sustainable, connected, and resilient urban future. This research lays the groundwork for future studies and policy initiatives that seek to harness the potential of advanced connectivity technologies for the betterment of cities and the planet. The findings underscore the imperative for a collective and concerted effort to build smart cities that prioritize sustainability, innovation, and inclusivity in the face of evolving urban challenges. This study significantly contributes to the field of smart city sustainability by providing a comprehensive understanding of the role 5G Green Networks in fostering environmental responsibility and technological advancement. The identified challenges, solutions, and policy implications offer actionable insights for governments, industry stakeholders, and researchers aiming to drive sustainable smart city initiatives.

REFERENCES

[1]   A. Caragliu, C. Del Bo, and P. Nijkamp, "Smart cities in Europe. Research Memoranda Series 0048 (VU University Amsterdam, Faculty of Economics, Business Administration and Econometrics)", *J. Urban Technol.*, vol. 18, no. 2, pp. 65-82, 2009.
        [http://dx.doi.org/10.1080/10630732.2011.601117]

[2]   T. Beatley, Ed., *Green cities of Europe: Global lessons on green urbanism.* Island press: Washington, DC, 2012.
        [http://dx.doi.org/10.5822/978-1-61091-175-7]

[3]   I. Jennings, and P. Newman, *Cities as sustainable ecosystems. Principles and practices,* 2008.

[4]   L. Zhang, H. Zhao, S. Hou, Z. Zhao, H. Xu, X. Wu, Q. Wu, and R. Zhang, "A survey on 5G millimeter wave communications for UAV-assisted wireless networks", *IEEE Access,* vol. 7, pp. 117460-117504, 2019.
        [http://dx.doi.org/10.1109/ACCESS.2019.2929241]

[5]   A. Agarwal, G. Misra, S. Agarwal, and K. Ghosh, "5G wireless cellular networks: a conceptual analysis on perception, network requirements and enabling technologies. J Inst Eng (India)", *Ser B.,* vol. 100, pp. 187-191, 2019.

[6]   F.A. Almalki, S.H. Alsamhi, R. Sahal, J. Hassan, A. Hawbani, N.S. Rajput, A. Saif, J. Morgan, and J. Breslin, "Green IoT for eco-friendly and sustainable smart cities: future directions and opportunities", *Mob. Netw. Appl.,* vol. 28, no. 1, pp. 178-202, 2023.
        [http://dx.doi.org/10.1007/s11036-021-01790-w]

[7]   P. Mishra, and G. Singh, "Energy management systems in sustainable smart cities based on the internet of energy: A technical review", *Energies,* vol. 16, no. 19, p. 6903, 2023.
        [http://dx.doi.org/10.3390/en16196903]

[8]   M.J. Shehab, I. Kassem, A.A. Kutty, M. Kucukvar, N. Onat, and T. Khattab, "5G networks towards smart and sustainable cities: A review of recent developments, applications and future perspectives", *IEEE Access,* vol. 10, pp. 2987-3006, 2022.
        [http://dx.doi.org/10.1109/ACCESS.2021.3139436]

[9]   A. Rehman, K. Haseeb, T. Saba, J. Lloret, and Z. Ahmed, "Mobility support 5G architecture with real-time routing for sustainable smart cities", *Sustainability (Basel),* vol. 13, no. 16, p. 9092, 2021.
        [http://dx.doi.org/10.3390/su13169092]

[10]   S.H. Alsamhi, F. Afghah, R. Sahal, A. Hawbani, M.A.A. Al-qaness, B. Lee, and M. Guizani, "Green internet of things using UAVs in B5G networks: A review of applications and strategies", *Ad Hoc Netw.*, vol. 117, p. 102505, 2021.
[http://dx.doi.org/10.1016/j.adhoc.2021.102505]

[11]   C. Yang, P. Liang, L. Fu, G. Cui, F. Huang, F. Teng, and Y.A. Bangash, "Using 5G in smart cities: A systematic mapping study", *Intelligent Systems with Applications,* vol. 14, p. 200065, 2022.
[http://dx.doi.org/10.1016/j.iswa.2022.200065]

[12]   N.I. Sarkar, and S. Gul, "Green computing and internet of things for smart cities: technologies, challenges, and implementation", In: *Green Computing in Smart Cities. Simulation and Techniques,* 2021, pp. 35-50.
[http://dx.doi.org/10.1007/978-3-030-48141-4_3]

[13]   M. Shehab, T. Khattab, M. Kucukvar, and D. Trinchero, "The role of 5G/6G networks in building sustainable and energy-efficient smart cities", In: *7ᵗʰ International Energy Conference (ENERGYCON)*, 2022, pp. 1-7.

[14]   D. Pradhan, H.M. Tun, and A.K. Dash, "IoT : Security & Challenges of 5G Network in Smart Cities", *Asian Journal of Convergence in Technology,* vol. 8, no. 2, pp. 45-50, 2022.
[http://dx.doi.org/10.33130/AJCT.2022v08i02.010]

[15]   P. He, N. Almasifar, A. Mehbodniya, D. Javaheri, and J.L. Webber, "Towards green smart cities using Internet of Things and optimization algorithms: A systematic and bibliometric review", *Sustainable Computing: Informatics and Systems,* vol. 36, p. 100822, 2022.
[http://dx.doi.org/10.1016/j.suscom.2022.100822]

[16]   A. Oad, H.G. Ahmad, M.S.H. Talpur, C. Zhao, and A. Pervez, "Green smart grid predictive analysis to integrate sustainable energy of emerging V2G in smart city technologies", *Optik (Stuttg.),* vol. 272, p. 170146, 2023.
[http://dx.doi.org/10.1016/j.ijleo.2022.170146]

[17]   A. Khelifi, O. Aziz, M.S. Farooq, A. Abid, and F. Bukhari, "Social and economic contribution of 5G and blockchain with green computing: Taxonomy, challenges, and opportunities", *IEEE Access,* vol. 9, pp. 69082-69099, 2021.
[http://dx.doi.org/10.1109/ACCESS.2021.3075642]

[18]   G.F. Huseien, and K.W. Shah, "Potential applications of 5G network technology for climate change control: A scoping review of Singapore", *Sustainability (Basel),* vol. 13, no. 17, p. 9720, 2021.
[http://dx.doi.org/10.3390/su13179720]

[19]   D. Pradhan, P.K. Sahu, A. Dash, and H.M. Tun, "Sustainability of 5G green network toward D2D communication with RF-energy techniques", *International Conference on Intelligent Technologies (CONIT),* pp. 1-10, 2021.
[http://dx.doi.org/10.1109/CONIT51480.2021.9498298]

[20]   S. Painuly, S. Sharma, and P. Matta, "Future trends and challenges in next generation smart application of 5G-IoT", In: *5ᵗʰ International Conference on Computing Methodologies and Communication (ICCMC),* 2021.

[21]   N. Gupta, P.K. Juneja, S. Sharma, and U. Garg, "Future Aspect of 5G-IoT Architecture in Smart Healthcare System", In: *5ᵗʰ International Conference on Intelligent Computing and Control Systems (ICICCS),* 2021.

[22]   M. Lu, G. Fu, N.B. Osman, and U. Konbr, "Green energy harvesting strategies on edge-based urban computing in sustainable internet of things", *Sustain Cities Soc.,* vol. 75, no. 103349, p. 103349, 2021.
[http://dx.doi.org/10.1016/j.scs.2021.103349]

[23]   A. Heidari, N.J. Navimipour, and M. Unal, "Applications of ML/DL in the management of smart cities and societies based on new trends in information technologies: A systematic literature review", *Sustain Cities Soc.,* vol. 85, p. 104089, 2022.

[http://dx.doi.org/10.1016/j.scs.2022.104089]

[24]   S. Rani, R.K. Mishra, M. Usman, A. Kataria, P. Kumar, P. Bhambri, and A.K. Mishra, "Amalgamation of advanced technologies for sustainable development of smart city environment: A review", *IEEE Access,* vol. 9, pp. 150060-150087, 2021. Available from: https://ieeexplore.ieee.org/abstract/document/9600866
[http://dx.doi.org/10.1109/ACCESS.2021.3125527]

[25]   A. Kasznar, A. Hammad, M. Najjar, E. Linhares Qualharini, K. Figueiredo, C. Soares, and A. Haddad, "Multiple dimensions of smart cities' infrastructure: A review", *Buildings,* vol. 11, no. 2, p. 73, 2021.
[http://dx.doi.org/10.3390/buildings11020073]

CHAPTER 2

The Effective Cost-Reduction Plan for Particle Swarm Optimization-Based Mobile Location Monitoring in 5G Communications

Prabhakar Rath[1,*], Smita Rani Parija[1] and **Kishan Gupta[1]**

[1] *Department of Electronics & Communication Engineering, C. V. Raman Global University, Bhubaneswar, Odisha-752054, India*

Abstract: The focus on cost reduction within mobile communication networks has become a key subject of attention due to its significant proportion of the overall cost utilization structure of information and communication technology (ICT). This research digs into the area of 5G networks, which include a heterogeneous mix of mega cells and small cells with a clear demarcation between data and control planes. The paper considers two categories of information or data. There are two categories of data flow or traffic: high-rate traffic for data and low-rate data congestion. Large-scale cellular base stations, or MBSs, are responsible for controlling and regulating signals in the conventional architecture for separation. In contrast, a small cell base station (SBS) controls data transmission at both low and high rates. An MBS manages control signals and- the pace of data flow within the modified separation architecture under consideration, whereas an SBS controls a high-speed data flow. An efficient energy-saving method is presented to improve the cost-effectiveness of base stations (BSs). The amount of user equipment (UEs) seeking high-rate data traffic and the number of UEs present within overlapping areas that are generally covered by the considered BS and neighboring BSs are used to establish the operational state of a BS. To implement this cost-cutting method, Particle swarm optimization (PSO) finds an application to create a problem related to optimizing something and find its answer. The findings unequivocally demonstrate that the suggested energy-saving approach, as implemented within the redesigned split network design, surpasses the energy efficiency achieved by traditional energy-efficient techniques, Both of them have distinct network structures that are basic and customized. Additionally, this suggested plan significantly reduces cumulative latency, offering a highly promising strategy for enhancing overall network efficiency.

Keywords: Cellular mobile systems, Location management, Particle swarm optimization (PSO), 5G networks.

---

[*] **Corresponding author Prabhakar Rath:** Department of Electronics & Communication Engineering, C. V. Raman Global University, Bhubaneswar, Odisha-752054, India; E-mail: prabhakar@cvrp.edu.in

Devasis Pradhan, Mangesh M. Ghonge, Nitin S. Goje, Alessandro Bruno and Rajeswari (Eds.)

INTRODUCTION

An essential concern associated with the issue of global warming is the steadily rising amount of energy that is consumed [1], and diminishing cost utilization. The use of mobile networks for communication has drawn attention because it accounts for a significant portion of the whole cost of technology for information and communication (ICT). The mobile communication network cost will be more crucial in the future with increasing traffic load in the upcoming networks with 5G [2, 3]. According to reference [4], in mobile communication networks, a base station (BS) is the main source of cost usage, and the BS's cost utilization is influenced by traffic load, which varies according to the geographical location.

Significant effort has been made to increase the energy savings of a BS to lower the cost use of mobile communication networks. The primary idea of work to lower the cost usage of a BS is to turn off as many BS components as feasible once they are no longer required [4]. For instance, the Base Station (BS) can enter a slumber mode through deactivating power to most of its components when they are not in active use. As depicted in Fig. (**1**), if the BS experiences minimal traffic and neighboring BSs can adequately cover the traffic, it can be powered off to achieve significant energy savings [5].

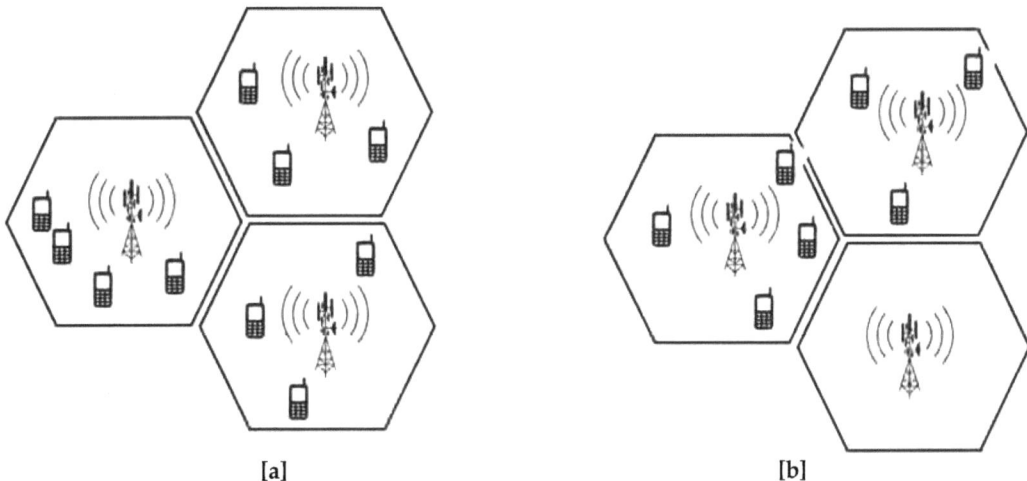

[a]   [b]

Fig. (1). (Base station (BS) activation and deactivations). (**a**) Activation; (**b**) deactivation [1].

Contextual information, such as dynamic changes in demand and channel for traffic conditions within a setting at which the Base Station (BS) and User Equipment (UE) connections are ever-changing, is used to determine the operational status of the BS. In a prior study [6], two schemes known as "greedy-on" and "greedy-off" were introduced. These schemes change the BS state to either on or off. It was found that the "greedy" strategy demonstrated improved energy efficiency provided certain conditions were met. In another study [7], various methods for reducing the number of active BSs were presented, and a traffic-intensity-aware multi-cell cooperation scheme was introduced. In this scheme, the BS state is adjusted to the off state based on its traffic density, which is categorized into peak-hour traffic and off-peak traffic, depending on the UE's traffic demand. Furthermore, coverage gaps are addressed by neighboring BSs that remain in one state. The cell zooming approach is also presented [8] as an alternative to the BS activation and deactivation procedures [4, 7], which adjusts the coverage area of BSs to allow cells to zoom in and out in response to changes in traffic load and channel conditions, as shown in Fig. (2).

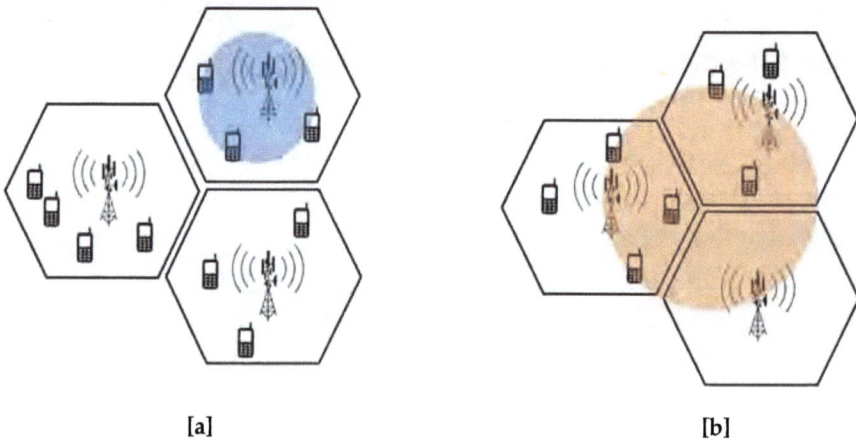

[a]   [b]

Fig. (2). Cell zooming can be done in two ways: (**a**) zooming in; (**b**) zooming out [2].

A BS that is busy and experiencing high traffic loads will zoom in, while nearby BSs will zoom away to prevent any possible coverage gaps. A BS zooms out while nearby BSs zoom in when they are not busy and have little traffic. Cell zooming allows the Base Station (BS) that is zoomed in to go into a sleep state to save energy and lower operating expenses. In this case, the region that the dormant BS had formerly serviced is covered by nearby BSs.By reducing the cost associated with active base stations, the cell zooming system aims to balance traffic distribution to enhance the system's overall energy effectiveness. To oversee the data coming from the switched-off BSs, active BSs must fill up any coverage gaps and work with neighboring BSs [9].

Additionally, expanding on the initial idea of cell zooming, detailed in [10], the authors introduce an efficient cost-saving approach for BSs. This approach involves deactivating BSs during periods of very low traffic load while allowing active BSs to expand their cell coverage by increasing transmission power. This strategy contributes to energy and cost savings in response to fluctuating traffic demands. Three cell zooming strategies were provided by some authors [11]: continuous, discrete, and fuzzy. To cover its farthest user through continual cell zooming, a base station (BS) modulates broadcast power. The discrete cell zooming approach allows for only discrete transmit power values. Lastly, in the fuzzy cell zooming technique, a BS transmit power is increased slightly from each discrete level as an extension of the discrete system. In a study [12], the authors presented a two-tier cellular network energy-saving small cell zooming strategy in which small cell base stations (BSs) dynamically choose whether to switch on, turn off, zoom in, and zoom out in response to changing traffic loads, UE speeds, and UE locations while ensuring UE data is satisfied.

Efforts to conserve energy in Base Stations (BSs) while efficiently serving User Equipments (UEs) have become a focal point in recent research [13 - 15]. Although shutting down BSs can result in significant energy savings, it introduces a considerable delay in responding to users' traffic demands. This is because BSs need to be reactivated from Measures to cater to UEs, a process that requires a fair amount of time. To address the challenge of prolonged reactivation times for deactivated BSs, an alternative approach has been explored. Instead of completely turning off all components, BSs deactivate most of their components to conserve energy and swiftly respond to users' demands by entering a sleep state, as opposed to being entirely switched off. A prior study [13] classified the battery states as on, standby, sleep, and off according to how active it is and how much energy the BS consumed. Since all BS operations are fully operational, the on-state has the highest power consumption.

A Base Station (BS) that is in the standby state can swiftly switch to the on state when necessary because only a few parts of the system are not in use. These parts include the temperature-compensated crystal oscillator (TCXO) heater and radio frequency (RF). While it uses more energy than a BS in the sleep or off states, a backup battery in BS standby mode uses less energy compared to the one that is active. A BS uses even less power when it's in the sleep state than when it is in the on or standby state; this has a resemblance to the mode with low power covered [14]. A base station (BS) only turns on necessary parts while it is in sleep mode, such as the CPU core, backend connection, and power supply. As a result, it takes longer to get from the sleep state to the on state than it does to go from the ready condition to the on state. A BS should be in the sleep state if its traffic load is relatively low. Nonetheless, the on-state is preferred if there is an excessive amount of traffic. Since all of a BS's operations are inactive when it is off, it uses no power at all. The change from an off to an on state takes the longest. After analyzing the connection between a BS's energy efficiency and traffic load, the authors [15] found that the level of sleep was more productive under low traffic loads. The ideal threshold value for going into a sleep state was also established by the authors. The authors in another study [16] looked into the connection between the length of residence during sleep and electricity efficiency.

Initiatives to save BS energy *via* innovative network architecture have been made in tandem with research on energy efficiency in BS states. In a typical heterogeneous network architecture, one primary cell is topped by several smaller cells. Small cells are made to accommodate the growing volume of data traffic and shifting traffic distribution. However, as each base station (BS) in the macro cell and small cell must manage data traffic in addition to the control signal, energy efficiency has a little effect on a typical heterogeneous network. This means that BS vigilance must be maintained at all times. Unused small Base Stations (BSs) inside a diverse network architecture overlapping both tiny and large cell BSs are usually turned off to save energy [17]. In an earlier work [18], a tiny cell BS used the user equipment's (UEs) locations and traffic load to regulate the small cell base station's activation and deactivation. However, this method introduced high signaling overhead.

A different method described in a study [19] uses distributed activation and deactivation systems, in which a tiny cell base station (BS) is not in use. When it is not servicing any user equipment (UEs), it reawakens regularly to monitor UE activity. To maximize the number of active and inactive small cell base stations, the DANCE techniques, or Device-Assisted-Networking for Cellular Greening [20] concentrate on activating or deactivating BSs for tiny cells based on fluctuating traffic loads. Within a different study [21], the authors suggested an algorithm that maximizes energy efficiency by simultaneously controlling the user associations and BSs' on/off statuses. Furthermore [22], a study determines the tiny cell's on/off state base stations (BSs) by utilizing two distinct uniform and non-uniform UE distributions. These tactics seek to achieve a balance between efficient network operation and energy conservation.

Researchers have looked at how energy-efficient Base Stations (BSs) within diverse networks are shared with several proprietors in several studies [23 - 25]. To save power, the creators [23] use distributed game theory. The idea is for UEs from one mobile network operator to be supplied by another, enabling the former operator's BSs to be turned off when not in use. A cooperative BS switching-off technique is also presented [24]. This approach allows BSs to be deactivated when their traffic load is low and allows BSs controlled by other mobile network carriers to effectively cover the UEs they service. The authors [25] suggest a game that takes user associations and price decisions into account. The goal of this game is to maximize network efficiency while taking into account the financial aspects of roaming agreements between operators. It is according to the connection between roaming charges and user affiliations. A revolutionary concept involving the separation of control and user planes, or a "control split plane and user plane," was presented to further improve the energy efficiency of Base Stations (BSs) [26, 27]. As shown in Fig. (**3**), a heterogeneous network design with this separation of data and control is managed by a macro cell BS (MBS), which is responsible for controlling the data traffic and control signals. On the other hand, since the MBS supplies the control signals, the small cell BS (SBS) is responsible only for managing data traffic.When there is little traffic in the areas that SBSs cover, they can use a lot less electricity. This is because the macro cell continuously provides control signals, thus they do not need to stay active to support them. To save energy, SBSs can be turned into an inactive mode. Heterogeneous networks provide the capacity to manage the constantly growing data traffic while also lowering Base Station (BS) energy usage [28]. However, when data traffic density is high or a BS's transmission power is low, the configuration of Small Cell BSs (SBSs) affects the performance of heterogeneous networks.

Moreover, to support User Equipment (UEs) from either inactive small cells or macro cells, SBSs could need to be active, which could result in higher energy consumption [28, 29]. SBSs are outfitted with high transmission power in the context of a cloud-cooperated heterogeneous network architecture, particularly in the direction of heavily trafficked locations, which improves data transmission efficiency [30]. The deployment of cloud radio access networks, or C-RANs, further minimizes the use of needless resources [31], a study that covers the technical issues related to BS toggle switch on/off strategies in 5G networks, presents the state of the art in terms of open subjects, technological issues, and BS on/off switching. A traffic-based dynamic BS switching on/off technique is proposed [32], specifically for a separated network design that separates traffic BSs from coverage BSs. This strategy incorporates a haphazard sleeping schedule. The authors [33] provide repulsive and random strategies in a network architecture that is divided. Each BS is turned off in the random system with a specific probability, represented by letter "p." Conversely, in the repulsive scheme, small cell BSs located less than "R" away from a macro cell BS are disabled.

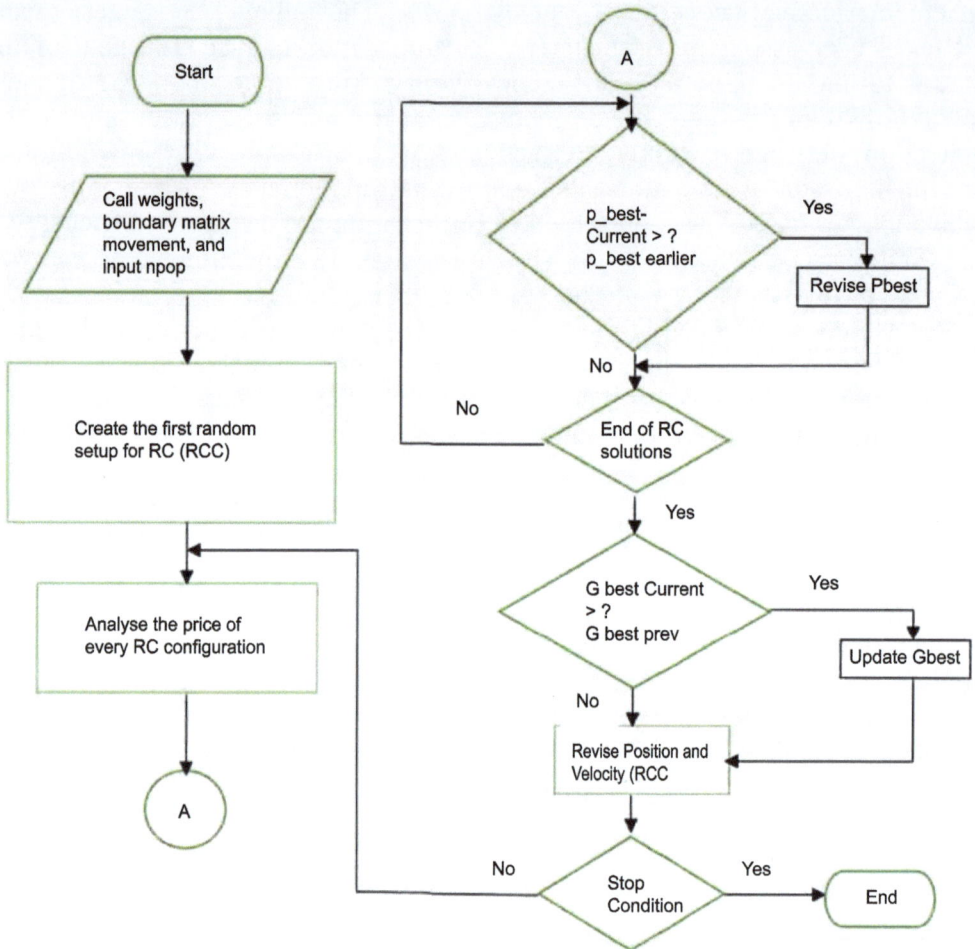

Fig. (3). PSO algorithm flowchart.

In recent studies [34 - 38], a great deal of research effort has gone into creating algorithms and optimization models for the administration and design of 5G networks. The main findings of this field's research are outlined below: a 5G network architecture based on reusable functional blocks (RFBs), which addresses the optimization challenge of dynamically managing RFBs [34]. The objective is to optimize user throughput or reduce the number of active nodes while taking into account limitations associated with user data, RFB location, 5G node capacity, and user coverage. Network function virtualization (NFV) infrastructure power consumption can be reduced by using a robust mixed-integer optimization approach [35]. This approach deals with virtual network functions' (VNFs') erratic resource demands. To reduce the significant computing time required to solve the robust optimization problem, the authors present a quick triangular heuristic known as " placement of robust and green VNFs" [36] to reduce the energy usage of network infrastructure and processing in 5G networks. We use both heuristic and exact approaches, focusing on how quickly the suggested heuristic converges when compared to exact solvers. In a study [37], the authors leverage virtual functions to develop and construct a video streaming service that makes use of 5G network capabilities for mobile edge computing [38]. A study describes an effective Particle Swarm Optimization (PSO)-based technique that uses RFBs to send users of 5G networks high-definition video content. Either increasing user throughput or reducing the number of 5G nodes that are actively in operation should be the main goals.

Numerous aspects of 5G network architecture and management are the subject of these research projects, which offer helpful insights and workable solutions to raise network efficiency and performance. In the field of wireless networks, wireless network design is considered fundamental, and for a deeper understanding, readers are directed to the following works [39 - 42]. The optimization concerns in telecommunications are covered in detail [39], whereas models and optimization issues for cellular network design are covered in another study [40]. Current problems with wireless network architecture in the modern era are also presented in a study [41].

A comprehensive overview of the application applying mathematical optimization concepts and methods to the architecture of wireless networks are given in a study [42], along with a full analysis of a hierarchy of design challenges. These research projects explore several aspects of the architecture and management of 5G networks, providing insightful analysis and workable solutions to improve network efficiency and performance. Network architecture is among the most significant fields in wireless technology; those who like to learn more are encouraged to look through the publications listed below [40 - 42]. A study explores the wide range of optimization problems in the telecom sector [39]. Models and optimization issues about the design of cellular networks are also examined [40].

The difficulties that are now being faced by wireless network designers are presented in a study [41]. In addition to a thorough explanation of a hierarchy of design difficulties [42], a study offers a thorough overview of applying theories of mathematical optimization and methods in wireless connection architecture. A thorough understanding of the optimization and design challenges in wireless networks is provided by these references.

The multi-objective nature of wireless network design is a major issue covered in some studies [53 - 55]. WCDMA network planning is approached as a multi-objective problem by the authors [53], whose objectives are to maximize system capacity and reduce installation costs. For green network planning of single-frequency network-based orthogonal frequency division multiplexing (OFDM) schemes, the authors [54] present a multi-objective problem with the objective functions of coverage optimization, transmitter minimization, exposure minimization, energy efficiency, carbon footprint, and green deployment. To simultaneously increase the number of clients serviced and decrease the supplied utility, the authors [55] developed a unique optimization model for resource assignment in heterogeneous wireless networks. With the release of numerous important studies, the subject of data uncertainty in wireless network design and resilient wireless network architecture has grown in prominence [56 - 62]. The authors [55] present a novel optimization model for resource assignment in heterogeneous wireless networks that seek to maximize the number of concurrently served consumers while minimizing the given utility.

Along with the publication of several important studies, the subject of uncertainty in data when designing wireless networks and the resilient design of wireless networks has gained prominence [56 - 62]. Below is a synopsis of these works: Robust optimization is utilized [56] to tackle traffic unpredictability in the design of telecommunication networks, with a particular emphasis on design issues of capacitated networks. The robust counterpart of a linear program with an unclear matrix of a coefficient is created [57], especially as soon as the uncertainty set is multi-band [58]. The ambiguity of the jamming issue is tackled by introducing a resilient cutting-plane method that optimizes the arrangement and positioning of jammers. This technique is influenced by multiband robust optimization. To handle traffic unpredictability in a multiperiod network design problem [59], a study offers a solid optimization approach. To find answers, the authors suggest a hybrid approach that combines exact large neighborhood search with ant colony optimization. For wireless local area networks to conserve energy [60], resilient optimization is suggested that takes user mobility and rate unpredictability into consideration. The writers [61, 66 - 68] concentrate on the best way to distribute transmitter power in a wireless network with random interference and useful link coefficients [62 - 65]. Some researchers develop a stochastic programming model for channel allocation and offer a yield management technique to optimize income under demand uncertainty. Base Station (BS) energy reductions in both homogeneous and heterogeneous networks have been the subject of research; most of these studies, however, have not taken segregated control and data planes into account. The majority of research on 5G networks' base stations' (BSs) energy efficiency has focused on modeling and performance analysis for networks with distinct control and data planes. As far as the authors are aware, not much research has been done on an effective technique to save energy in base stations (BSs) in 5G networks that have separate control and data planes [33]. Most evaluations of the suggested protocol do not take mobility into account [68 - 70].

The authors of this research propose an energy-saving approach for 5G networks with independent control and data planes. The condition of a BS is ascertained by counting the number of User Equipment (UEs) seeking high-rate data traffic and the number of UEs within overlapping zones covered by the surrounding BSs and the considered BS. For this energy-saving plan, they define an optimization issue and use particle swarm optimization to find answers. The suggested protocol significantly expands on earlier work [34] by providing comprehensive state management of BS algorithms, a comprehensive formulation of optimization problems, a useful application of particle swarm optimization, and a large number of numerical examples derived from simulations.

The paper is organized as follows: Part 2 outlines the suggested cost-saving plan. In Section 3, particle swarm optimization is used to examine the scheme's performance. In Section 4, numerical examples are given and explored in detail. Section 5 concludes by summarizing the research and outlining possible next paths. Previously all the papers suggested only energy-saving issues but our paper provides detailed information about the cost-effective optimization of mobile location management in 5G communications.

OUTLINE OF THE SUGGESTED COST-SAVING PLAN

Reporting cell planning (RCP) is a prominent method for location management difficulties as suggested in a study [61]. Under RCP, a portion of the cellular network's cells are called reporting cells, and they frequently broadcast short messages to clarify their function. The location of a mobile terminal (MT) is updated only upon entering a new reporting cell. The position update is only available in the most recent reporting unit's surrounding cells not submitting reports upon the arrival of a mobile user's call discussed in Table **1**. Stated differently, like this collection among cells that report acts as a gate, where calls are routed and mobile terminal positions are updated [62]. This project aims to optimize the RCP setup for the lowest feasible total price of paging and location updates. One tool for worldwide optimization is particle swarm optimization, and it is seen as a discrete optimization issue discussed in Table **2**. The issue with RCP is now an NP-complete problem as a result [63].

Table 1. Test network call arrival and weights during movement.

Cell	*Wco*	*Wmo*	Cell	*Wco*	*Wmo*
1.	714	1039	9.	546	1829
2.	120	1476	10.	221	296
3.	414	262	11.	856	793
4.	639	442	12.	652	317
5.	419	1052	13.	238	507
6.	332	1902	14.	964	603
7.	494	444	15.	789	1479
8.	810	1103	16.	457	756

Table 2. Real network call arrival and weights during movement.

Cell	*Wco*	*Wmo*	Cell	*Wco*	*Wmo*
1.	349.00	807.00	9.	36.00	163.00
2.	61.00	700.00	10.	289.00	497.00
3.	321.00	398.00	11.	2,045.00	3,567.00
4.	21.00	185.00	12.	187.00	1,917.00
5.	6.00	1,197.00	13.	38.00	289.00
6.	30.00	980.00	14.	105.00	236.00
7.	285.00	533.00	15.	193.00	451.00
8.	324.00	1,556.00	16.	480.00	700.00

As a result, the overall location management cost associated with a specific RCP setup is calculated as follows:

$$TotalCost = \propto \times \sum_{i \in R} w_{mo} + \sum_{j=0}^{N} w_{co} \times v(j) \tag{1}$$

Here, we considered a 4x4 cellular network. The cost per call arrival to total cost is obtained as follows:

$$\text{Arrival N} = \frac{Total\ Cost\ Cost\ per\ Call}{wcj\ j} \tag{2}$$

Particle Swarm Optimization is used to Examine the Scheme's Performance

A population-based strategy that is iterative is Particle Swarm Optimization. With each iteration, starting with RC setups as binary vectors, a population of solutions (particles) is created, and the solutions are developed into better ones. The particle movement of each iteration is determined by global best, or the best solution discovered overall (g_best), and found thus far For p_best. The goal of our optimization task is to find the optimal reporting cell configuration by minimizing the total cost. The steps below outline the PSO approach for the RCP problem:

Step 1: First, we generate a population of particles, or solutions, for the given RCP issue. A reporting cell configuration represents a single solution. A reporting cell is designated with a '1', whereas a non-reporting cell is designated with a '0'.

Step 2: Use equation (4) to determine the cost for each particle (P).

Step 3: Save p-best and g-best for every particle during the loop. If the cost of the p-best is lower now than it was earlier, the current p-best should be used instead of the previous one. In a similar vein, the current g-best ought to take the place of the old one if the current g-best's cost is less.

Step 4: Use that following equations to calculate the particle's velocity and position change.

$$Vnext = w * Vcurrent + c1 * r1 * [p\_bestcurrent - Pcurrent] + c2 * r2 * [g\_bestcurrent- \\ Pcurrent] \tag{3}$$

The particle is relocated from $P_{current}$ to P_{next}, using the equation,

$$Pnext = Pcurrent + Vnext \tag{4}$$

Step 5: Continue from Step 2 until the termination requirement is satisfied, where the weighting parameters are w, c1*r1, and c2*r2.

The PSO algorithm flow chart is displayed in Fig. (**3**).

NUMERICAL EXAMPLES ARE GIVEN AND EXPLORED IN DETAIL

The simulation results are verified for real-world networks and applied to reporting cell planning benchmark tasks [62, 63]. Several experiments are conducted in this part using MATLAB for 4x4 networks, an Intel Core (TM) 2 Duo processor running at 2.20 GHz, 4GB of RAM, and a CPU 32-bit operating system. The gathered data is presented together with the analysis and conclusions that go along with it. The data is collected during three different iterations: fifty, one hundred, and five hundred. The iteration count is set as a stop condition. There are 150 people in the simulation. The issue is applied here for even and odd networks of different sizes. In symmetric networks, experiments are conducted (4x4(16 cells)). The output of the simulation displays the convergence diagrams of the top-performing /optimal cost of the RCP setup for every iteration.

Function of Fitness Importance

That fitness function is utilized in this challenge of reporting cell planning to calculate each location's management expense, as given by Eq (1). This means that a cellular network made up of registered both non-reporting and reporting cells is created for every possible solution that is presented, using the fitness value as a guide.

Characterizing Parameters

Since it is the foundation for algorithm development, the first definition of parameters is one of the most important steps. The initial population of potential solutions is matched by the quantity of particles. Every individual particle is a size N binary resolution vector, which is the total number of RCP network cells. The value of the binary configuration in the RCP network establishes whether a cell is a reporting or non-reporting cell. The parameters of the PSO algorithm must first be defined as follows: It is assumed that the weighting factors w, c1*r1, and c2*r2 are 1.5, 1, and 0.7, respectively. The constant =10 [10] denotes the proportion of location registration expenses to paging expenses.

Fig. (4) displays the plot of the confluence of the overall expense of location management for the 4x4 test networks. The 4x4 real network graphs of convergence are shown in Fig. (5). The experimental results of the higher network 8x8 test and the realistic network are then shown in Figs. (4 and 7), respectively. The convergence map for the odd test network with a 6x6 size in a similar vein, displays its combination with the practical 6x6 real network. The convergence graphs show that as the number of iterations grows, convergence slows down from the initial rapid pace. As a result, the simulation is run initially with a lower iteration number of 150 and later with higher values of 250. This is because PSO [64 - 67] starts with costly, inefficient, and haphazardly generated solutions. The price of the algorithm's particles gets closer to one another as the number of iterations increases, which causes the convergence to improve results slowly.

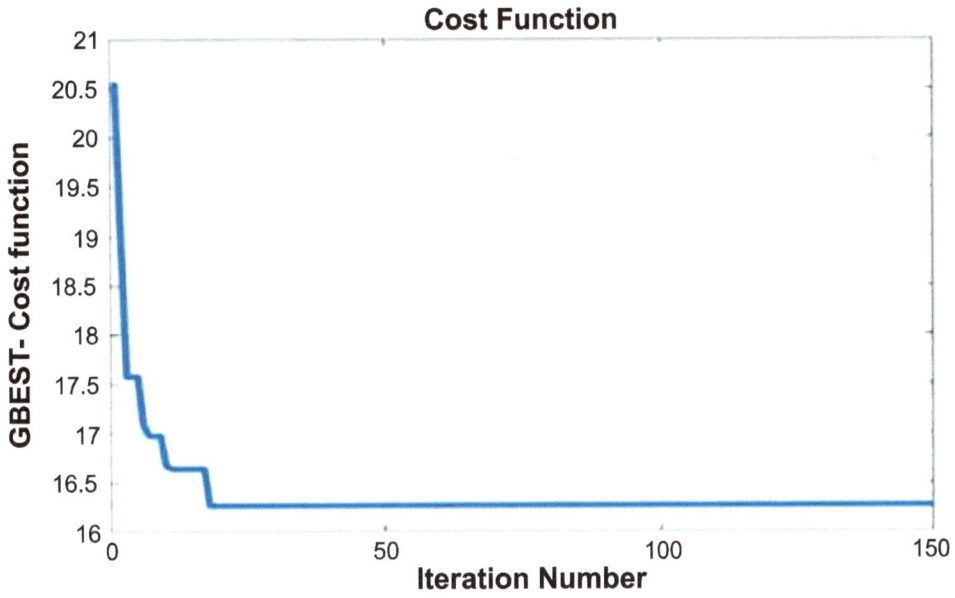

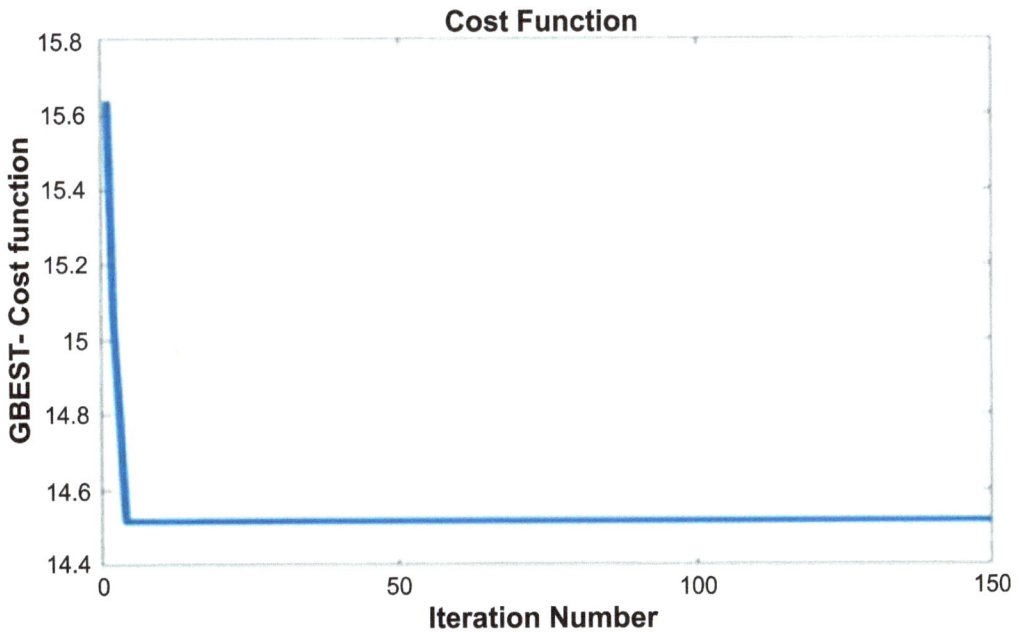

Fig. (4). Outcomes of 150, 250, 4x4 test data network simulation repetition.

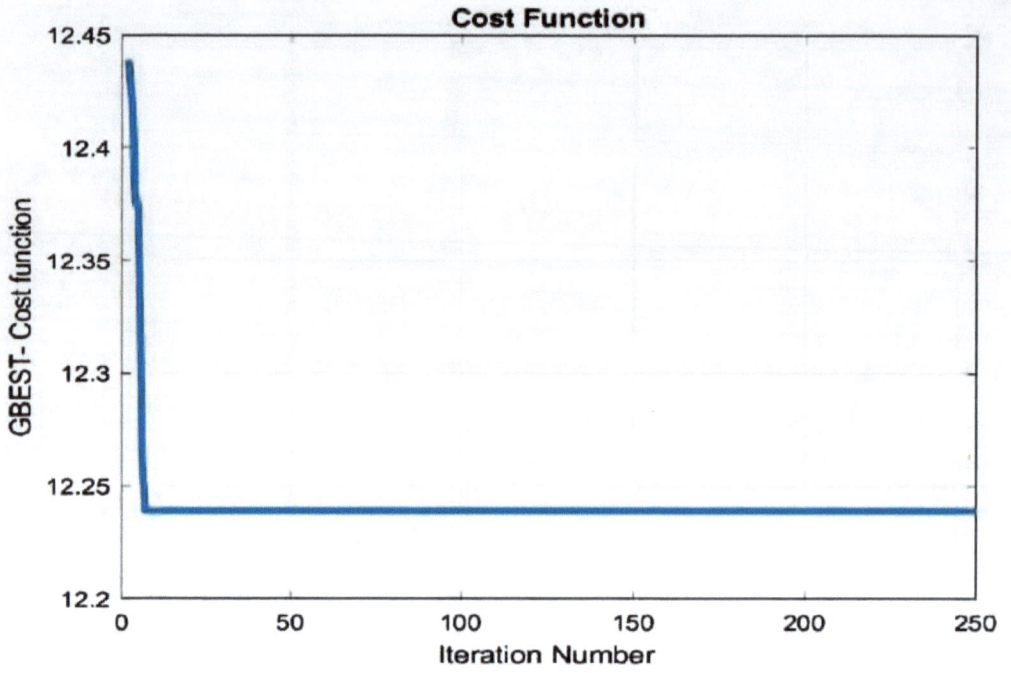

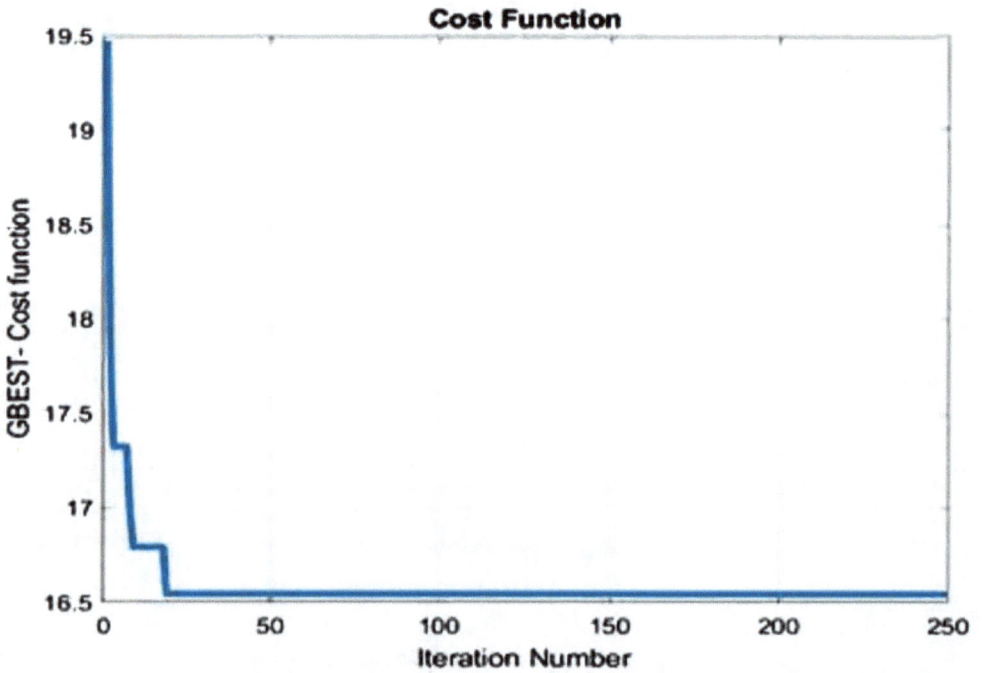

Fig. (5). Outcomes of 150, 250, 4x4 real data network simulation repetition.

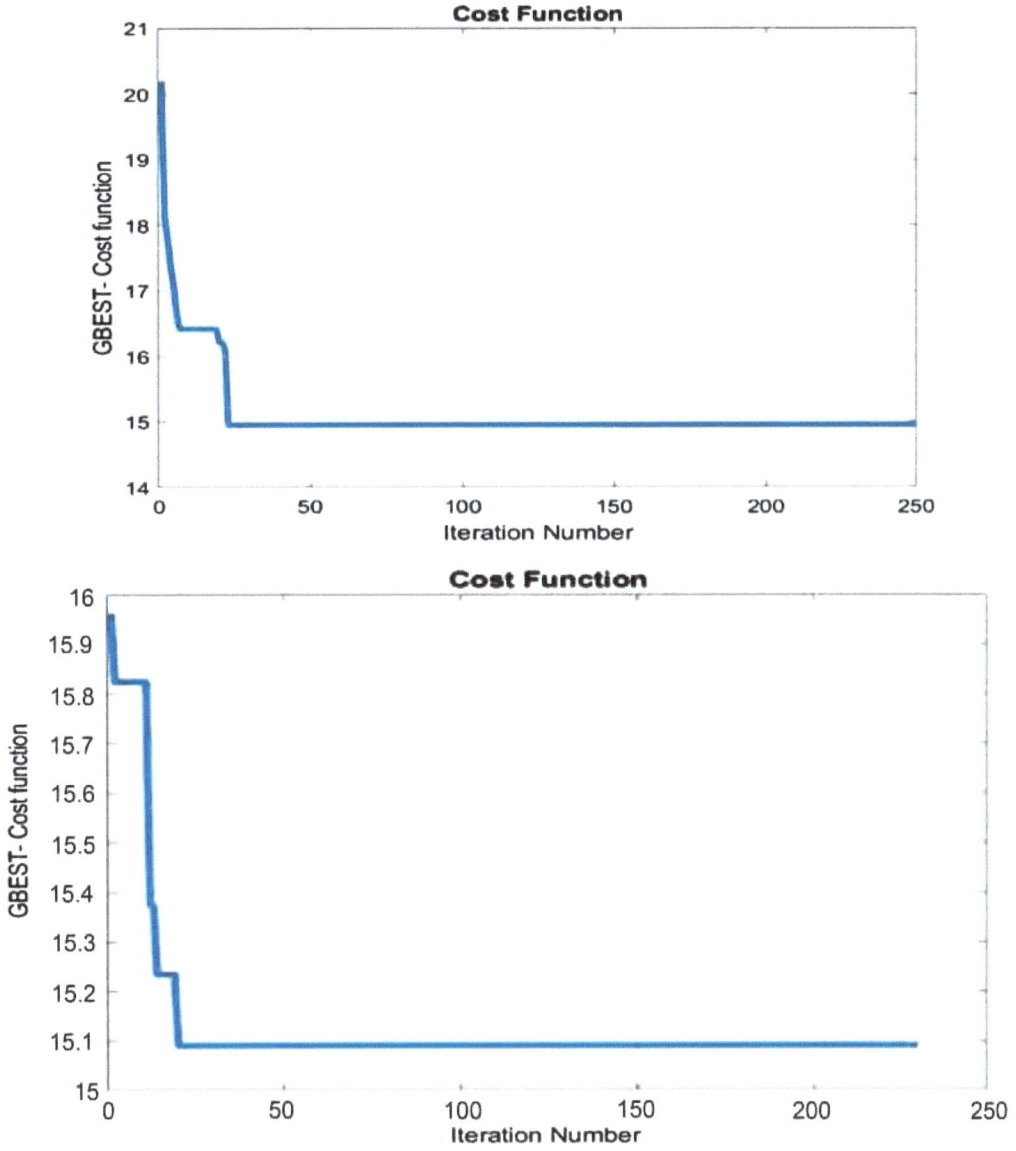

Fig. (6). Outcomes of 150, 250, 8x8 real data network simulation repetition.

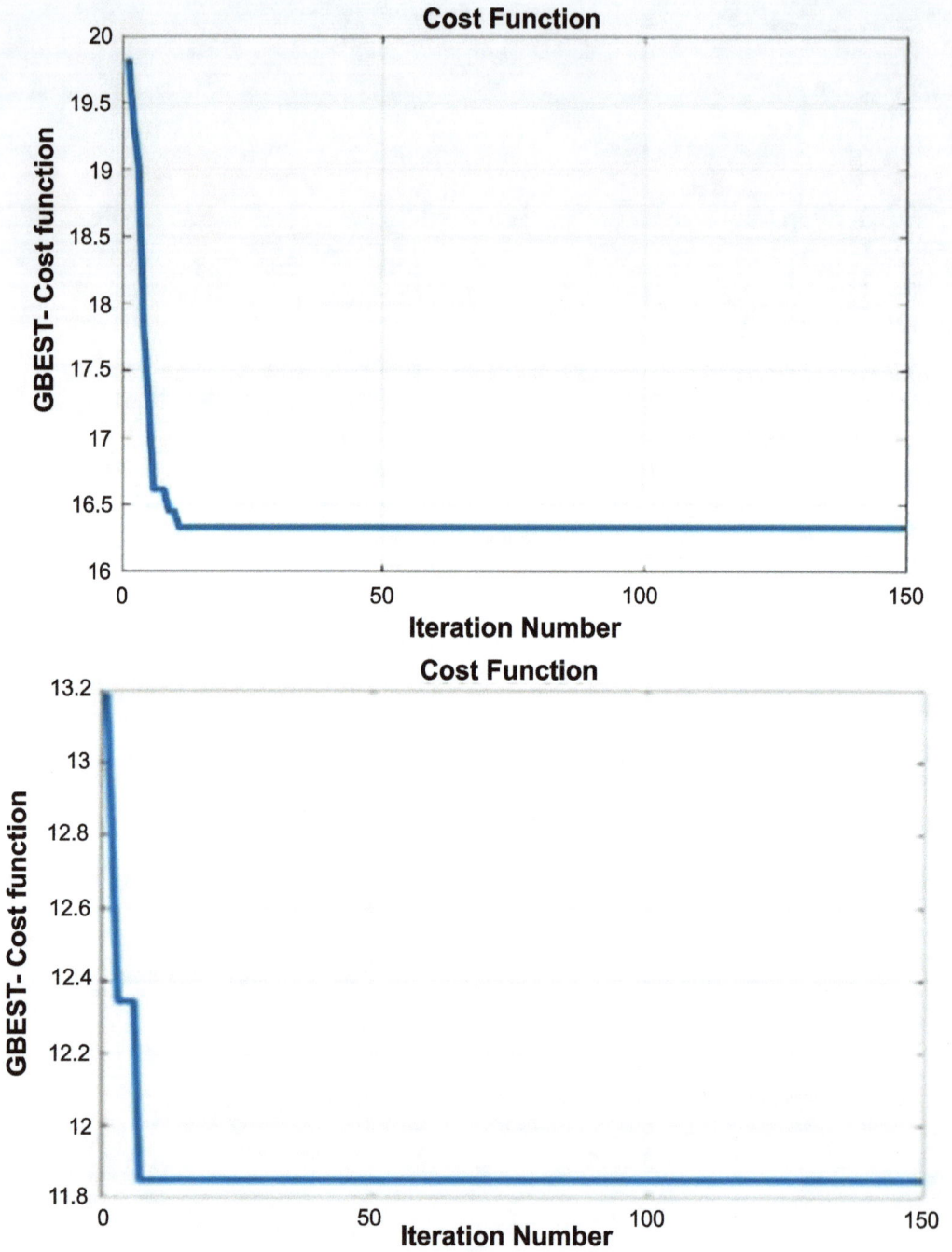

Fig. (7). Outcomes of 150, 250, 6x6 real data network simulation repetition.

CONCLUSION BY SUMMARIZING THE RESEARCH AND OUTLINING POSSIBLE NEXT PATHS

In this paper, an RCP-based location management technique is provided. The particle swarm optimization approach is used to minimize the location management cost function, which is the definition of the optimization issue. The ideal group of cells for reporting is discovered regarding a 16-cell system, also known as a 4x4 cellular network that is employed as a model for showing issue formulation and solution technique. Tables **2**, **3**, **4**, and **5** explore and explain the expanded 4x4, 8x8, and 6x6 simulation findings, as well as their verification for the identical network with actual data. Subsequent investigations will concentrate on utilizing hybrid optimization techniques, broadening the scope to practical networks, and adapting the optimization issue to any cellular network with a size of N. It will also employ more thorough comparison analyses with other metaheuristic approaches. It is necessary to investigate dynamic mobile location management. Mobile location management is crucial given the exponential increase in mobile subscribers and the technological advancements from 5G and beyond. This function must be implemented in a way that is more effective and optimized for mobile networks of the future.

Table 3. Cost per call arrival analysis using PSO for different networks of size 4x4, 8x8, and 6x6.

Reference Network	BPSO	Real Network	BPSO
4x4	87162	4x4	58089
8x8	479211	8x8	570481
6x6	669108	6x6	644044

Table 4. Total location management cost analysis using PSO for different networks of size 4x4, 8x8, and 6x6.

Ref Data Network	Repetition	Standard Deviation	Mean	Median	Minimum	Maximum
4x4	150	1.4804	7.0199	7.7933	7.7933	7.1332
4x4	250	1.0848	7.9344	7.7933	7.7933	8.0228
8x8	150	2.5844	6.4024	6.0208	6.0208	7.6361
8x8	250	1.6361	6.2600	6.0208	6.0208	8.3536
6x6	150	1.8770	7.3874	7.0938	7.0938	8.9092
6x6	250	1.3430	7.3126	7.0938	7.0938	7.3920
4x4	150	1.0069	9.9937	9.8063	9.8060	10.2753
4x4	250	0.3891	9.8635	9.8063	9.8063	9.0713
8x8	150	0.3961	9.8394	9.8063	9.8063	9.6746
8x8	250	0.6914	9.8950	9.8063	9.8063	10.0298

(Table 4) cont.....

Ref Data Network	Repetition	Standard Deviation	Mean	Median	Minimum	Maximum
6x6	150	0.9067	9.9968	9.8063	9.8063	9.3036
6x6	250	0.1596	9.8198	.8063	9.8063	9.2302

Table 5. Comparing the cost per call arrival using PSO to previous research.

Reference Network	Proposed Algorithm BPSO	BPSO	DE	MHN	ACO	TS	GA
-	-	[11]	[10]	[9]	[7]	[8]	[6]
4x4	3.5197	NA	12.47	NA	11.252	11.242	12.242
8x8	5.4508	12.782	13.68	N/A	13.725	14.525	13.782
6x6	8.4502	NA	NA	NA	NA	NA	NA

REFERENCES

[1]   S.R. Parija, P.K. Sahu, and S.S. Singh, "Cost reduction in location management using reporting cell planning and particle swarm optimization", *Wireless Pers. Commun.,* vol. 96, no. 1, pp. 1613-1633, 2017.
[http://dx.doi.org/10.4018/978-1-5225-2322-2.ch008]

[2]   S. Parija, S. Swayamsiddha, P.K. Sahu, and S.S. Singh, "Profile-based location update for cellular network using mobile phone data", *Microsyst. Technol.,* vol. 27, no. 2, pp. 369-377, 2021.
[http://dx.doi.org/10.1007/s00542-019-04367-6]

[3]   S.R. Parija, N.P. Nath, P.K. Sahu, and S.S. Singh, "Dynamic intelligent paging in mobile telecommunication network", *Sadhana,* vol. 43, no. 2, 2018.
[http://dx.doi.org/10.1007/s12046-018-0804-3]

[4]   S. Swayamsiddha, S. Parija, S. Sudhansu, and P. Singh, "Reporting cell planning-based cellular mobility management using a binary artificial bat algorithm", *Heliyon,* vol. 5, p. 1276, .
[http://dx.doi.org/10.1016/j.heliyon.2019.e01276]

[5]   S.R. Parija, S.S. Singh, and P.K. Sahu, "Reporting cell planning for cost reduction using binary genetic algorithm", *International Journal of Convergence Computing,* vol. 2, no. 3/4, p. 235, 2016.
[http://dx.doi.org/10.1504/IJCONVC.2016.10010694]

[6]   Z. Niu, Y. Wu, J. Gong, and Z. Yang, "Cell zooming for cost-efficient green cellular networks", *IEEE Commun. Mag.,* vol. 48, no. 11, pp. 74-79, 2010.
[http://dx.doi.org/10.1109/MCOM.2010.5621970]

[7]   L.T. Berger, A. Schwager, and J.B. Hubaux, *Routing and Scheduling in Wireless Networks: Theory and Algorithms.* 1st ed. Cambridge University Press, 2014.

[8]   J. Zhang, C. Yue, H. Li, and Y. Wu, "A hybrid algorithm for optimization in wireless sensor networks", *Sensors,* vol. 18, no. 8, 2018.
[http://dx.doi.org/10.3390/s18082495]

[9]   Xiaobing Wu, Guihai Chen, and S.K. Das, "Avoiding energy holes in wireless sensor networks with nonuniform node distribution", *IEEE Trans. Parallel Distrib. Syst.,* vol. 19, no. 5, pp. 710-720, 2008.
[http://dx.doi.org/10.1109/TPDS.2007.70770]

[10] H. Choi, S. Park, and S. Lee, "Energy-efficient clustering algorithms for wireless sensor networks", *Wirel. Commun. Mob. Comput.,* vol. 16, no. 8, pp. 1006-1020, 2016.

[11] J. Kennedy, and R. Eberhart, "Particle swarm optimization", *Proc. IEEE Int. Conf. Neural Networks,*

pp. 1942-1948, 1995.
[http://dx.doi.org/10.1109/ICNN.1995.488968]

[12]   R. Storn, and K. Price, "Differential evolution – A simple and efficient heuristic for global optimization over continuous spaces", *J. Glob. Optim.,* vol. 11, no. 4, pp. 341-359, 1997.
[http://dx.doi.org/10.1023/A:1008202821328]

[13]   C. Blum, and X. Li, "Swarm intelligence in optimization", In: *Swarm Intelligence: Introduction and Applications.* 1st ed. Springer: Berlin, Germany, 2008, pp. 43-85.
[http://dx.doi.org/10.1007/978-3-540-74089-6_2]

[14]   M. Dorigo, and L.M. Gambardella, "Ant colony system: a cooperative learning approach to the traveling salesman problem", *IEEE Trans. Evol. Comput.,* vol. 1, no. 1, pp. 53-66, 1997.
[http://dx.doi.org/10.1109/4235.585892]

[15]   K. Deb, "Multi-objective optimization using evolutionary algorithms", *Wiley,* 2001.

[16]   S. Kirkpatrick, C.D. Gelatt Jr, and M.P. Vecchi, "Optimization by simulated annealing", *Science,* vol. 220, no. 4598, pp. 671-680, 1983.
[http://dx.doi.org/10.1126/science.220.4598.671] [PMID: 17813860]

[17]   Z. Michalewicz, *Genetic Algorithms + Data Structures = Evolution Programs.* 3rd ed. Springer-Verlag, 1996.
[http://dx.doi.org/10.1007/978-3-662-03315-9]

[18]   F. Glover, "Tabu search: A tutorial", *Interfaces,* vol. 20, no. 4, pp. 74-94, 1990.
[http://dx.doi.org/10.1287/inte.20.4.74]

[19]   D. Karaboga, and B. Akay, "A comparative study of Artificial Bee Colony algorithm", *Appl. Math. Comput.,* vol. 214, no. 1, pp. 108-132, 2009.
[http://dx.doi.org/10.1016/j.amc.2009.03.090]

[20]   E. Bonabeau, M. Dorigo, and G. Theraulaz, *Swarm Intelligence: From Natural to Artificial Systems.* Oxford University Press, 1999.
[http://dx.doi.org/10.1093/oso/9780195131581.001.0001]

[21]   S. Mirjalili, "Genetic algorithm", In: *Evolutionary Algorithms and Neural Networks: Theory and Applications.* Springer, 2019, pp. 43-62.
[http://dx.doi.org/10.1007/978-3-319-93025-1_4]

[22]   A.E. Eiben, and J.E. Smith, *Introduction to Evolutionary Computing.* Springer, 2003.
[http://dx.doi.org/10.1007/978-3-662-05094-1]

[23]   J.H. Holland, *Adaptation in Natural and Artificial Systems.* University of Michigan Press, 1975.

[24]   S.J. Russell, and P. Norvig, *Artificial Intelligence: A Modern Approach.* 4th ed. Pearson, 2020.

[25]   I.F. Akyildiz, W. Su, Y. Sankarasubramaniam, and E. Cayirci, "Wireless sensor networks: a survey", *Comput. Netw.,* vol. 38, no. 4, pp. 393-422, 2002.
[http://dx.doi.org/10.1016/S1389-1286(01)00302-4]

[26]   D. Culler, D. Estrin, and M. Srivastava, "Overview of sensor networks", *Computer,* vol. 37, no. 8, pp. 41-49, 2004.
[http://dx.doi.org/10.1109/MC.2004.93]

[27]   S. Misra, S.D. Raghunathan, and V.C. Prasad, *Guide to Wireless Sensor Networks.* Springer, 2009.
[http://dx.doi.org/10.1007/978-1-84882-218-4]

[28]   K. Akkaya, and M. Younis, "A survey on routing protocols for wireless sensor networks", *Ad Hoc Netw.,* vol. 3, no. 3, pp. 325-349, 2005.
[http://dx.doi.org/10.1016/j.adhoc.2003.09.010]

[29]   A. Ephremides, "Energy concerns in wireless networks", *IEEE Wirel. Commun.,* vol. 9, no. 4, pp. 48-59, 2002.

[http://dx.doi.org/10.1109/MWC.2002.1028877]

[30]   Y. Yao, and J. Gehrke, "The cougar approach to in-network query processing in sensor networks", *SIGMOD Rec.,* vol. 31, no. 3, pp. 9-18, 2002.
[http://dx.doi.org/10.1145/601858.601861]

[31]   C. Intanagonwiwat, R. Govindan, and D. Estrin, "Directed diffusion: A scalable and robust communication paradigm for sensor networks", *Proc. ACM MOBICOM,* pp. 56-67, 2000.
[http://dx.doi.org/10.1145/345910.345920]

[32]   W. Heinzelman, A. Chandrakasan, and H. Balakrishnan, "Energy-efficient communication protocol for wireless microsensor networks", *Proc. 33rd Annu. Hawaii Int. Conf. Syst. Sci.,* pp. 1-10, 2000.
[http://dx.doi.org/10.1109/HICSS.2000.926982]

[33]   S. Lindsey, and C.S. Raghavendra, "PEGASIS: Power-efficient gathering in sensor information systems", *Proc. IEEE Aerospace Conf.,* pp. 1125-1130, 2002.
[http://dx.doi.org/10.1109/AERO.2002.1035242]

[34]   O. Younis, and S. Fahmy, "HEED: a hybrid, energy-efficient, distributed clustering approach for ad hoc sensor networks", *IEEE Trans. Mobile Comput.,* vol. 3, no. 4, pp. 366-379, 2004.
[http://dx.doi.org/10.1109/TMC.2004.41]

[35]   J.N. Al-Karaki, and A.E. Kamal, "Routing techniques in wireless sensor networks: a survey", *IEEE Wirel. Commun.,* vol. 11, no. 6, pp. 6-28, 2004.
[http://dx.doi.org/10.1109/MWC.2004.1368893]

[36]   H. Karl, and A. Willig, *Protocols and Architectures for Wireless Sensor Networks.* Wiley, 2005.
[http://dx.doi.org/10.1002/0470095121]

[37]   M. Perillo, and W. Heinzelman, "Providing application-aware quality of service in wireless sensor networks", *IEEE Commun. Mag.,* vol. 41, no. 2, pp. 140-147, 2003.

[38]   S.R. Madden, M.J. Franklin, J.M. Hellerstein, and W. Hong, "TinyDB: an acquisitional query processing system for sensor networks", *ACM Trans. Database Syst.,* vol. 30, no. 1, pp. 122-173, 2005.
[http://dx.doi.org/10.1145/1061318.1061322]

[39]   L. Kumar, M.P. Singh, and M.K. Ghose, "Energy-efficient clustering and routing algorithms for wireless sensor networks: A survey", *IJCSNS,* vol. 8, no. 7, pp. 250-260, 2008.

[40]   J. Polastre, R. Szewczyk, and D. Culler, "Telos: Enabling ultra-low power wireless research", *Proc. IPSN 2005,* pp. 364-369, 2005.

[41]   A. Manjeshwar, and D.P. Agrawal, "TEEN: A routing protocol for enhanced efficiency in wireless sensor networks", *Proc. IPDPS 2001 Workshops,* pp. 2009-2015, 2001.
[http://dx.doi.org/10.1109/IPDPS.2001.925197]

[42]   B. Krishnamachari, D. Estrin, and S. Wicker, "The impact of data aggregation in wireless sensor networks", *Proc. ICDCS 2002 Workshops,* pp. 575-578, 2002.
[http://dx.doi.org/10.1109/ICDCSW.2002.1030829]

[43]   M. Haenggi, *Opportunities and Challenges in Wireless Sensor Networks.* Springer, 2005.

[44]   M.H. Yaghmaee, and D. Adjeroh, "A new priority-based congestion control protocol for wireless multimedia sensor networks", *Proc. IEEE ICC 2008,* pp. 248-252, 2008.
[http://dx.doi.org/10.1109/WOWMOM.2008.4594816]

[45]   I. Akyildiz, T. Melodia, and K. Chowdury, "Wireless multimedia sensor networks: A survey", *IEEE Wirel. Commun.,* vol. 14, no. 6, pp. 32-39, 2007.
[http://dx.doi.org/10.1109/MWC.2007.4407225]

[46]   T. Rappaport, *Wireless Communications: Principles and Practice.* 2nd ed. Prentice Hall, 2002.

[47]   Y. Wang, and M. Vuran, "Spatial correlation-based collaborative medium access control in wireless

sensor networks", *IEEE Trans. Mobile Comput.,* vol. 8, no. 5, pp. 662-675, 2009.

[48] J.L. Hill, and D.E. Culler, "Mica: a wireless platform for deeply embedded networks", *IEEE Micro,* vol. 22, no. 6, pp. 12-24, 2002.
[http://dx.doi.org/10.1109/MM.2002.1134340]

[49] J.A. Stankovic, "Research challenges for wireless sensor networks", *ACM SIGBED Review,* vol. 1, no. 2, pp. 9-12, 2004.
[http://dx.doi.org/10.1145/1121776.1121780]

[50] K. Romer, and F. Mattern, "The design space of wireless sensor networks", *IEEE Wirel. Commun.,* vol. 11, no. 6, pp. 54-61, 2004.
[http://dx.doi.org/10.1109/MWC.2004.1368897]

[51] P.S.M.D.S.A. Pradeep, and C.V. Jawahar, "Energy-efficient routing techniques for wireless sensor networks: A survey", *Ad Hoc Netw.,* vol. 6, no. 7, pp. 1019-1027, 2008.

[52] Y. Xu, J. Heidemann, and D.W. Estrin, "Geography-informed energy conservation for ad hoc routing", *Proc. ACM/IEEE MobiCom,* pp. 70-84, 2001. Rome, Italy.
[http://dx.doi.org/10.1145/381677.381685]

[53] A. Boukerche, "Algorithms and protocols for wireless, mobile ad hoc networks", *Springer Science & Business Media,* 2008.
[http://dx.doi.org/10.1002/9780470396384]

[54] W.B. Heinzelman, A. Chandrakasan, and H. Balakrishnan, "Energy-efficient communication protocol for wireless microsensor networks", *Proc. 33rd Hawaii Int. Conf. Syst. Sci.,* pp. 1-10, 2000.
[http://dx.doi.org/10.1109/HICSS.2000.926982]

[55] S. Misra, I. Woungang, and S.S. Iyengar, *Guide to Wireless Sensor Networks.* Springer, 2009.
[http://dx.doi.org/10.1007/978-1-84882-218-4]

[56] P.S.L.M.A.B.M.A.M. Farooq, "Clustering techniques for wireless sensor networks: A survey", *Comput. Netw.,* vol. 53, no. 6, pp. 893-908, 2009.

[57] M.A. Khan, and H.T. Mouftah, "A survey of energy efficient routing protocols for wireless sensor networks", *IEEE Commun. Mag.,* vol. 51, no. 1, pp. 1-9, 2013.

[58] B. Karp, and H.T. Kung, "GPSR: Greedy perimeter stateless routing for wireless networks", *Proc. ACM MobiCom,* pp. 243-254, 2000.
[http://dx.doi.org/10.1145/345910.345953]

[59] M. Dorigo, and T. Stützle, *Ant Colony Optimization.* MIT Press, 2004.
[http://dx.doi.org/10.7551/mitpress/1290.001.0001]

[60] J.B. Schmitt, M.R.A.K. Tiwari, and W.W.M. Thomas, "Wireless sensor networks: A survey", *J. Comput. Sci. Technol.,* vol. 24, no. 1, pp. 1-15, 2009.

[61] F.B. Bastani, H.M.M.Y. He, and S. Xu, "Energy-efficient clustering in wireless sensor networks: A survey", *Comput. Commun.,* vol. 41, pp. 26-39, 2014.

[62] C.S. Raghavendra, S.S. Iyengar, and A.M.K. Manjeshwar, "Low-energy adaptive clustering hierarchy (LEACH) in wireless sensor networks", *IEEE Trans. Mobile Comput.,* vol. 2, no. 4, pp. 481-489, 2003.

[63] M. Yang, H. Xie, L. Xie, and D. Zhao, "Energy-efficient clustering schemes for large-scale wireless sensor networks", *J. Netw. Comput. Appl.,* vol. 34, no. 1, pp. 56-67, 2011.

[64] H.H. Chen, and K.H. Kim, *Mobile ad hoc networks: Research, theory, and applications.* Springer, 2009.

[65] M. R. B. and A. S. D. Krishnan, "A survey on energy efficient protocols for wireless sensor networks", *Comput. Commun.,* vol. 31, no. 7, pp. 1409-1420, 2008.

[66] A.S. Al-Hourani, S. Kandeepan, and H.S.S. Mohamed, "Energy efficient routing protocols for wireless

sensor networks", *Proc. ICCC 2015,* pp. 55-59, 2015.

[67]   I.F. Akyildiz, and Xudong Wang, "A survey on wireless mesh networks", *IEEE Commun. Mag.,* vol. 43, no. 9, pp. S23-S30, 2005.
[http://dx.doi.org/10.1109/MCOM.2005.1509968]

[68]   S. R. T. and B. S. M. Kumar, "A survey on routing techniques in wireless sensor networks", *Int. J. Comput. Appl.,* vol. 15, no. 6, pp. 40-45, 2011.

[69]   J. Li, C.Y. Wang, Z.J. Liang, and J.W. Yang, "A survey on wireless sensor networks", *Int. J. Comput. Appl.,* vol. 3, no. 4, pp. 98-106, 2007.

[70]   Y.W. Chen, "Design and implementation of a wireless sensor network system for monitoring and controlling", *Int. J. Comput. Appl.,* vol. 7, no. 9, pp. 87-94, 2008.

CHAPTER 3

Smart Cities with 5G and Edge Computing in 2030

Pushpendra Pal Singh[1,*] and **Rakesh Kumar Dixit**[1]

[1] *G.L. Bajaj Institute of Management, Greater Noida, India*

Abstract: The emergence of smart cities represents a paradigm shift in urban development, harnessing technological advancements to address pressing challenges and improve quality of life for citizens. As we look towards the year 2030, the convergence of 5G and edge computing technologies promises to revolutionize the landscape of urban environments, unlocking unprecedented levels of connectivity, efficiency, and sustainability. This paper explores the transformative potential of integrating 5G and edge computing in shaping the smart cities of the future. Firstly, it delves into the foundational principles of smart cities, emphasizing the need for interconnectedness, data-driven decision-making, and citizen-centric design. Building upon this framework, it examines the distinct capabilities offered by 5G networks, such as ultra-low latency, high bandwidth, and massive device connectivity, and elucidates how these attributes facilitate the proliferation of IoT devices, autonomous systems, and immersive experiences within urban contexts. Moreover, the paper discusses key challenges and considerations associated with the deployment of 5G and edge computing infrastructures in urban environments, such as cybersecurity risks, regulatory frameworks, and equitable access. It advocates for collaborative efforts among stakeholders, including governments, industries, and communities, to address these challenges and ensure the responsible and equitable implementation of smart city technologies.

Keywords: Autonomous vehicles, IoT (Internet of Things), Real-time data processing, Smart infrastructure, Smart cities 2030, 5G networks.

INTRODUCTION TO SECURITY IN SMART CITIES

In today's rapidly advancing era of 5G and the Internet of Things (IoT), where technology is evolving at an unprecedented pace, the concept of building secure cities has emerged as a topic of paramount importance [1 - 5]. With the advent of 5G, which promises lightning-fast connectivity and the ability to support a massive number of devices simultaneously, and the widespread integration of IoT devices into our daily lives, cities are presented with both incredible opportunities

---

[*] **Corresponding author Pushpendra Pal Singh:** G.L. Bajaj Institute of Management, Greater Noida, India;
E-mail: pps2907@gmail.com

Devasis Pradhan, Mangesh M. Ghonge, Nitin S. Goje, Alessandro Bruno and Rajeswari (Eds.)

and significant challenge [6 - 8]. As urban centres continue to evolve and progress, becoming increasingly interconnected and digitally integrated, it is evident that the potential benefits that can be derived from this transformation are truly immense. This paradigm shift holds the promise of enhancing efficiency, sustainability, and overall quality of life for the residents of these urban areas [9]. However, it is important to note that as technology continues to advance and become more integrated into our daily lives, it also brings with it a multitude of security challenges that cannot be ignored. These challenges must be addressed in a comprehensive manner to ensure the safety and privacy of citizens, the smooth functioning of critical infrastructure, and the overall resilience of the urban environment [10 - 12]. It is crucial that we take these challenges seriously and implement effective measures to mitigate any potential risks that may arise. By doing so, we can embrace the benefits of technological advancements while also safeguarding the well-being of our society [13 - 23].

Kho and Jeong [24] proposed a Hazard Analysis and Critical Control Points (HACCP)-based cooperative model for smart factories in South Korea. HACCP is a systematic approach to identifying and mitigating risks in manufacturing processes. This model integrates 5G connectivity to enable real-time monitoring and control, enhancing the safety and quality of products in smart manufacturing environments.

The seamless integration of 5G networks and Internet of Things (IoT) devices has revolutionised the way smart cities operate. By harnessing the power of these advanced technologies, cities are now able to collect and analyse massive volumes of data in real-time [25, 26]. This influx of data has paved the way for more intelligent decision-making processes across various sectors, including urban planning, traffic management, energy consumption, healthcare, and much more. With the advent of 5G networks, smart cities have gained unprecedented connectivity and speed, enabling them to handle the vast amounts of data generated by IoT devices [27]. These devices, ranging from sensors and cameras to smart metres and wearables, are embedded throughout the city's infrastructure, creating a network of interconnected devices that constantly gather and transmit data [28, 29]. One of the key benefits of this integration is the ability to enhance urban planning. By analysing real-time data on population density, traffic patterns, and environmental factors, city planners can make informed decisions to optimise the allocation of resources and improve the overall quality of life for residents. For example, they can identify areas with high traffic congestion and implement intelligent traffic management systems to alleviate the problem [30 - 32]. Moreover, the integration of 5G and IoT enables cities to monitor and manage energy consumption more efficiently. Smart metres installed in buildings can provide real-time data on energy usage, allowing for better energy management and the identification of areas where energy efficiency can be improved. This not only reduces costs but also contributes to As we delve deeper into the realm of connectivity, we find ourselves presented with a myriad of exciting opportunities that have the potential to revolutionise the way we live, work, and interact with the world around us [33, 34].

Temesvári *et al.* [35] provide a review of mobile communication technologies and their applications in the manufacturing sector. The paper discusses the role of 5G in supporting various manufacturing processes, including predictive maintenance, supply chain management, and flexible production. Understanding the impact of 5G on manufacturing is essential for industries seeking to leverage its benefits.

However, it is important to acknowledge that this newfound connectivity also brings with it a set of challenges and risks that we must navigate carefully. One of the foremost concerns that arise from this interconnectedness is the increased vulnerability to cyberattacks. With more devices and systems being connected to the internet, the potential for malicious actors to exploit vulnerabilities and gain unauthorised access to sensitive information becomes a pressing issue [36]. This can lead to devastating consequences, ranging from financial losses to compromised personal data and even threats to national security. Furthermore, the risk of data breaches looms large in this interconnected landscape. As more data is generated, stored In order to effectively establish secure cities within this ever-evolving landscape, it is imperative to adopt a comprehensive and multifaceted approach that encompasses a wide range of measures [37]. These measures should not only focus on technological advancements, but also take into consideration the organisational and regulatory aspects of city security. Technological measures play a crucial role in enhancing the security of cities. The implementation of advanced surveillance systems, such as high-definition cameras and facial recognition software, can significantly bolster the ability to monitor and detect potential threats. Additionally, the integration of smart technologies, such as Internet of Things (IoT) devices, can provide real-time data and analysis, enabling authorities to respond swiftly to any security breaches shown in Fig. (1). However, it is important to recognise that technology alone cannot guarantee the safety of a city.

These challenges encompass threats such as vulnerabilities introduced by network slicing, which is a key feature of 5G networks. The authors also propose innovative solutions, including enhanced authentication methods, to address these challenges. This source is valuable for gaining insights into the dynamic landscape of 5G security [38].

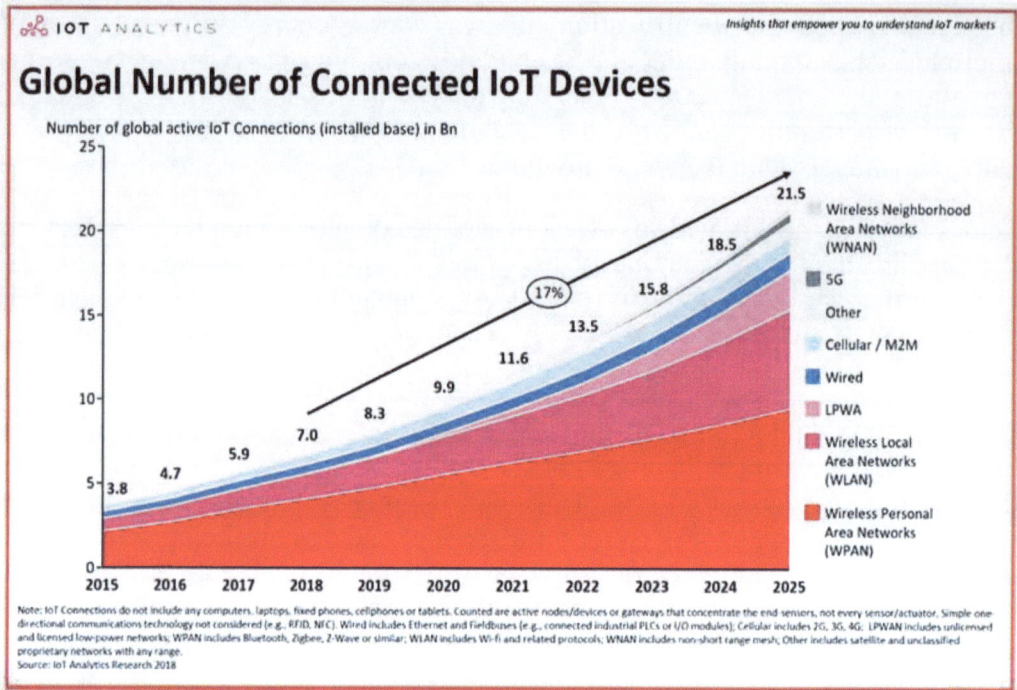

Fig. (1). Global number of IoT devices *vs* network type by IoT analytics [1].

Security Challenges in Modern Urban Environments

As urban areas embrace technological progress and shift towards becoming intelligent cities, they encounter a variety of security dilemmas that necessitate meticulous contemplation and alleviation. These predicaments arise from the amalgamation of 5G networks and Internet of Things (IoT) devices into urban infrastructure, introducing novel susceptibilities and plausible perils. Several of the primary security obstacles in contemporary urban settings encompass:

- **Cyberassaults and breaching:** The interlinked nature of intelligent urban systems generates countless access points for cyber offenders to capitalise on. Cyber intrusions, information breaches, and malware assaults can disturb vital services, jeopardise confidential data, and influence public reliance.
- **Data confidentiality concerns:** Intelligent urban systems accumulate extensive quantities of data from Internet of Things (IoT) devices and sensors. Safeguarding the confidentiality of citizen information becomes crucial, as unauthorised entry or mishandling of this data can result in identity theft, monitoring, and violations of individual privacy.
- **Network weaknesses:** The implementation of 5G networks facilitates rapid data transmission, but it also broadens the target area for potential intruders. Weaknesses in the network framework could be leveraged to disturb

communication, jeopardise device functionality, or even initiate synchronised assaults.

- **IoT device safety:** Numerous IoT devices utilised in intelligent urban areas may possess restricted security attributes because of financial and architectural limitations. These gadgets can be jeopardised and utilised as access points for wider assaults on the municipality's framework.
- **Absence of standardisation:** The varied assortment of IoT devices and technologies employed in intelligent cities can result in a deficiency of standardised security measures. This renders it difficult to execute uniform security measures and oversee the complete ecosystem efficiently.
- **Physical infrastructure vulnerabilities:** As metropolitan systems become progressively digitalized, the dependence on electronic infrastructure can render cities vulnerable to physical assaults or catastrophes that affect both electronic and tangible systems [39].

Integration of 5G and IoT in Smart City Infrastructure

The amalgamation of 5G networks and IoT devices in intelligent metropolis infrastructure offers unparalleled possibilities for enhancing urban existence, but it also introduces intricate security considerations. 5G's substantial bandwidth and minimal latency facilitate instantaneous data interchange, bolstering implementations such as self-driving cars, telemedicine, and intelligent power networks. Internet of Things (IoT) gadgets, like detectors and effectors, amplify the gathering and conveyance of information for enhanced decision-making and asset administration [40].

Nevertheless, this amalgamation further amplifies the possible repercussions of security breaches. The velocity and magnitude of data transmission amplify the vulnerability, and the interconnectedness of diverse devices and systems renders it imperative to guarantee the safety of each constituent in the ecosystem. Without adequate security measures, the advantages of 5G and IoT could be eclipsed by cyber hazards and compromised public confidence [41].

Importance of Ensuring Data Privacy, Cybersecurity, and Citizen Safety

Ensuring information confidentiality, cyber defence, and citizen security is of utmost importance for the prosperous acceptance and durability of intelligent urban projects. This significance is emphasised by various factors:

- **Confidence and citizen involvement:** Citizens require confidence that their personal information is managed securely and that their confidentiality is honoured. Constructing and upholding this confidence is crucial for public involvement and embrace of intelligent urban solutions.

- **Resilience of infrastructure:** Assaults on intelligent urban infrastructure can disturb vital amenities such as transport, power, and healthcare, affecting the welfare of inhabitants. Enforcing sturdy cybersecurity measures amplifies the tenacity of the metropolitan setting.
- **Legal and regulatory compliance:** Many jurisdictions have strict regulations regarding data privacy and cybersecurity. Noncompliance with these guidelines can lead to lawful ramifications and harm to the standing of the municipality's governance.
- **Economic impact:** A major security breach could have significant economic consequences, affecting businesses, tourism, and investment in the city.
- **Moral considerations:** Moral concerns emerge when citizen information is gathered without their awareness or utilised for intentions they have not approved of. Honouring data confidentiality is not solely a lawful necessity but also a moral duty.

In summary, the amalgamation of 5G and IoT in an intelligent metropolis framework harbours vast potential for enhancing urban existence, yet it also presents security hurdles that necessitate resolution. By giving precedence to data confidentiality, cyber defence, and public welfare, cities can establish a groundwork of confidence and durability that enables them to completely utilise the advantages of technological progress while protecting the welfare of their inhabitants.

Alaba *et al.* (2017) offer a comprehensive survey of the security landscape within the context of the Internet of Things (IoT). IoT introduces unique security challenges due to the sheer volume and diversity of connected devices. This source highlights the critical importance of implementing robust security measures, including end-to-end encryption and secure device management, to protect IoT ecosystems from potential threats [42].

THREAT LANDSCAPE IN SMART CITIES

The progressing peril panorama in intelligent metropolises presents an intricate and manifold quandary that originates from the amalgamation of 5G networks and Internet of Things (IoT) technologies. As metropolitan settings become progressively interconnected and information-based, they become susceptible to a wide array of hazards. Cyber assaults aiming at vital infrastructure, like power networks and transit systems, may disturb crucial amenities and undermine the operation of the metropolis. Internet of Things (IoT) gadgets, with their restricted security characteristics, are vulnerable to manipulation for botnets, monitoring, and data infringements. Furthermore, the colossal quantities of data gathered and conveyed by intelligent urban systems can transform into a profitable objective for cyber offenders aiming to pilfer confidential data or undermine citizen confidentiality. The swift tempo of technological advancement and the vast magnitude of interconnected devices make it difficult to recognise and alleviate weaknesses efficiently. Consequently, tackling the perilous environment in intelligent metropolises necessitates an all-encompassing strategy that includes information security measures, regulatory structures, collaboration between the public and private sectors, and ongoing surveillance to protect urban infrastructure, citizen welfare, and the general robustness of the city ecosystem [43].

Alhilal *et al.* (2020) delve into the emerging domain of distributed vehicular computing, which is poised to play a pivotal role in 5G networks. This concept has significant implications for autonomous vehicles and intelligent transportation systems, where real-time data processing and decision-making are essential. Understanding the challenges and opportunities associated with distributed vehicular computing is vital for comprehending the transformative potential of 5G in the automotive sector [44].

Overview of Potential Threats and Vulnerabilities in Smart Cities

Intelligent cities, distinguished by their incorporation of 5G networks and Internet of Things (IoT) gadgets, encounter a plethora of perils and susceptibilities that can undermine their functionality and jeopardise citizen safety. These obstacles emerge from the intricate interaction of digital systems and tangible infrastructure, generating possibilities for malevolent individuals to capitalise on vulnerabilities. A few of the main dangers and susceptibilities comprise:

Cyber assaults Focusing on Vital Infrastructure

Intelligent metropolises depend on interlinked systems to oversee utilities, transport, healthcare, and beyond. Cyber malefactors can aim at vital infrastructure to disturb indispensable amenities, inducing extensive pandemonium and conceivably jeopardising lives.

Data Breaches

The expansive compilation and conveyance of data within intelligent urban environments can allure hackers to pursue valuable data. Data breaches can reveal delicate citizen information, encompassing individual identifiers and monetary documents, resulting in identity theft and deceit.

Confidentiality Worries

The widespread implementation of IoT gadgets, like monitoring cameras and detectors, may violate individuals' confidentiality. Unpermitted entry to these gadgets can lead to intrusive monitoring and a violation of personal limits.

Cyberattacks Targeting Critical Infrastructure:

One remarkable instance of this peril is the Ukrainian Power Grid Assault in 2015. Cybercriminals, thought to be linked with a sovereign entity, remotely breached the power grid's command systems, resulting in a widespread power outage. This occurrence emphasised the possibility for cyber assaults to disturb vital infrastructure and exhibited the necessity for strong security measures to safeguard against such assaults.

Information Leaks: In 2019, Baltimore's Ransomware Assault witnessed the city's computer systems contaminated with ransomware, rendering vital services unreachable and causing extensive turmoil. Even though the assault was not directly linked to intelligent urban infrastructure, it emphasised the susceptibility of municipal systems to cyber dangers. The occurrence stimulated conversations regarding the significance of data backup, encoding, and proactive cybersecurity precautions.

Confidentiality Worries: Instances of unauthorized surveillance through compromised IoT devices have also emerged. For instance, the Ring Doorbell Breach Episodes implicated malevolent individuals acquiring entry to residents' Ring doorbell cameras and surveilling their actions. These occurrences emphasise the possible encroachment of confidentiality that can happen when IoT devices are not sufficiently fortified. Fig. (**2**) discuss about the three layers of IoT Devices Architecture.

These tangible occurrences underscore the immediacy of implementing resilient security measures in intelligent cities to alleviate risks and weaknesses. By preemptively tackling cybersecurity worries, establishing rigorous data-safeguarding protocols, and engaging stakeholders in the creation of secure infrastructure, intelligent cities can amplify their resilience and guarantee the welfare of their inhabitants in an ever more interconnected urban terrain.

ROLE OF 5G AND IOT IN SMART CITY SECURITY

The incorporation of 5G networks and Internet of Things (IoT) technology in intelligent cities has a significant influence on augmenting urban safety. With the extensive implementation of IoT devices, encompassing detectors and cameras, instantaneous data gathering and examination become achievable, empowering swift menace identification and reaction. The rapid and minimal-delay capabilities of 5G facilitate quick transmission of data, enabling instantaneous surveillance, intelligent video analytics, and enhanced emergency response. These technologies enable cities to oversee public areas and vital infrastructure effectively, automate security duties, and improve the deployment of emergency services. Nevertheless, although 5G and IoT present noteworthy advantages, obstacles such as data confidentiality, encoding, and cyber safety must be tackled to guarantee the soundness and durability of intelligent urban security systems [45].

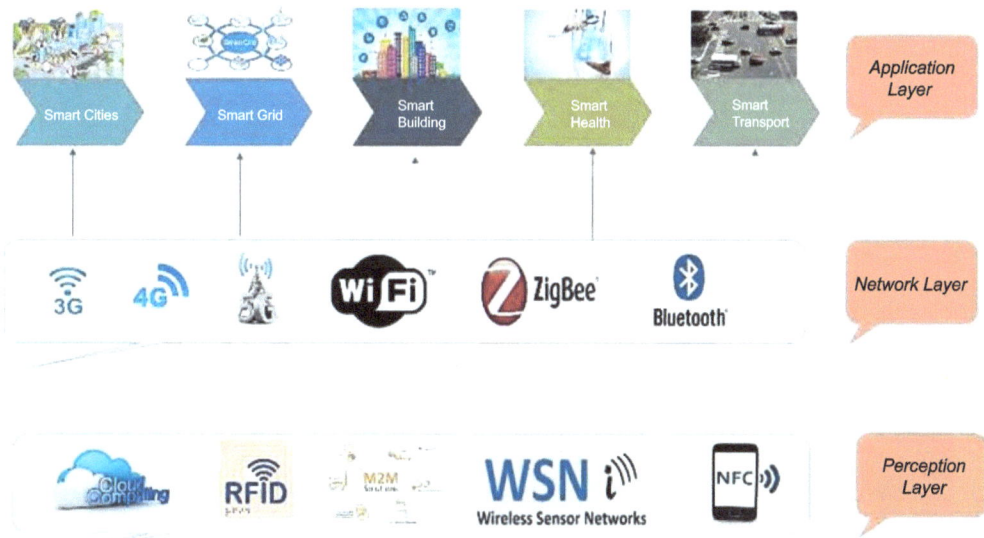

Fig. (2). IoT three-layer architecture [2].

How 5G and IoT Technologies Contribute to Enhanced Security Measures

The incorporation of 5G networks and Internet of Things (IoT) technologies brings a revolutionary change to how security measures are executed and supervised in intelligent cities shown in Fig. (**3**). These advancements not only amplify the effectiveness of current security measures but also empower the establishment of more forward-thinking, information-based, and adaptable security networks. Here is a comprehensive examination of how 5G and IoT contribute to improved security measures:

- **Live Data Collection and Analysis:** IoT devices, embedded with diverse sensors, seize and transmit live data from urban environments. These gadgets can oversee traffic flux, ecological circumstances, pedestrian motion, and beyond. 5G's rapid connectivity and minimal delay enable this information to be transmitted and examined nearly instantaneously. This live data gathering offers security personnel current insights into urban activities, enabling swift recognition of irregularities or possible dangers.

- **Prognostic Analytics and Menace Detection:** 5G-fueled IoT devices enable the execution of anticipatory analytics algorithms. By scrutinising past and up-to the-minute data patterns, these algorithms can anticipate plausible security menaces or occurrences. For example, atypical conduct in the collective motion could suggest a potential safety concern. Anticipatory analytics enable preemptive measures to be taken before a situation escalates, enhancing overall urban security.

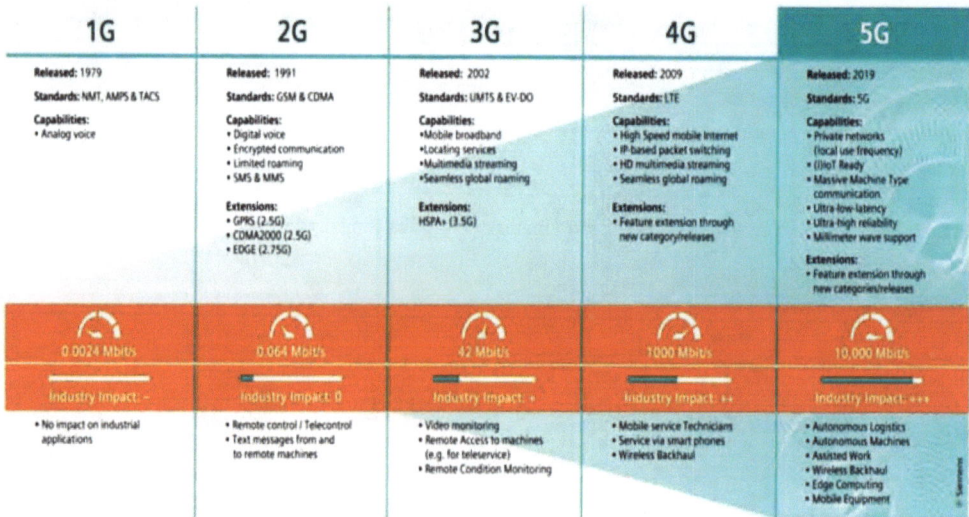

Fig. (3). First generation to fifth generation mobile network [4].

- **Extensive surveillance and observation:** IoT-enabled surveillance cameras and sensors can be strategically positioned throughout the city to oversee public areas, transportation networks, and vital infrastructure. With 5G's capacity, high-quality video streams can be transmitted effortlessly to surveillance centres. Security staff can remotely retrieve these streams, enabling them to promptly react to developing occurrences, whether it is a vehicular collision or a dubious behaviour.
- **Clever video analytics:** IoT-enabled cameras can surpass conventional surveillance by integrating clever video analytics. Artificial intelligence algorithms have the capability to recognise particular entities or actions in video

streams. This capacity enables automated surveillance and alarm generation when predetermined occurrences, such as unattended parcels or lingering, are identified. By reducing the requirement for human surveillance, security personnel can concentrate on more crucial responsibilities.

- **Enhanced emergency reaction:** 5G-enabled IoT devices contribute to enhanced emergency response systems. Location data from linked devices provides precise information about incidents' exact locations, enabling emergency services to react more swiftly. In circumstances where each moment is crucial, this can be vital for rescuing lives and reducing harm.
- **Distant manipulation and mechanisation:** 5G's minimal delay empowers instantaneous distant manipulation and mechanisation of security systems. For example, officials can remotely manage obstacles, entryways, and even unmanned aerial vehicles to promptly address emerging dangers or occurrences. Automation lessens the reliance on bodily presence while upholding security watchfulness.
- **Data integration for comprehensive perspectives:** IoT devices produce data from diverse origins, like congestion, climate, and online networks. 5G's capacities facilitate data amalgamation, empowering security personnel to acquire comprehensive understandings into urban activities. This extensive comprehension assists in recognising patterns, tendencies, and possible dangers that may not be evident when contemplating individual data sources.

In summary, the amalgamation of 5G and IoT technologies greatly enhances the productivity, efficacy, and promptness of security measures in intelligent urban areas. These technologies enable cities to utilise up-to-the-minute data, anticipatory analytics, smart video analytics, and mechanisation to establish more secure urban surroundings. While these progressions offer immense advantages, cautious contemplation of data confidentiality, encoding, and cyber safety is imperative to guarantee the reliance and robustness of intelligent urban security systems. (Li *et al.*, 2018).

Amine and Oumnad (2017) explore the profound impact of the Internet of Things (IoT) on various facets of human activities. From healthcare monitoring to smart homes and urban planning, IoT has the potential to revolutionize the way we live and interact with our surroundings. This source underscores the importance of IoT as a transformative technology and provides insights into the diverse applications of IoT across different domains.

Real-time Data Collection, Analysis, and Response Capabilities

The amalgamation of 5G networks and Internet of Things (IoT) technologies in intelligent cities brings forth revolutionary capabilities for instantaneous data gathering, examination, and reaction, transforming the manner in which metropolitan surroundings are supervised and safeguarded shown in Fig. **(4)**.

IoT Applications (Specific IoT Application along with Data Management, Data Mining, Business Support System, Operation Support System)	IoT Application Security (TLS 1.3, DTLS, HTTPS, Authentication & Authorization, Trust between Network and Services)
IoT Communication (3GPP Access-5G, 4G, CS Data, Non -3GPP Access - WiFi)	IoT Communication Security (5G network, Slicing, Hybrid Authentication, PKI, Mutual Authentication, E2E security, TLS 1.3, DTLS, OSCORE)
IoT Devices (Sensors, Actuators)	IoT Device Security (Secured Identity, Symmetric key cryptography)

Fig. (4). IoT layer architecture and IoT security [4, 5].

- **Live data gathering:** IoT devices, varying from detectors to monitoring cameras, are strategically positioned across intelligent cities, incessantly gathering assorted sets of data. These gadgets collect data on traffic trends, ecological circumstances, power usage, and additional details. With 5G's rapid connectivity, this data is conveyed to central command centres with negligible latency, guaranteeing that information is up-to-date and precise.
- **Swift data examination:** The accessibility of immense quantities of up-to-the minute information is solely significant if it can be examined promptly and effectively. 5G's minimal delay and ample bandwidth enable the swift transfer of information to cloud-based analytical platforms. Here, artificial intelligence algorithms can process and scrutinise data at unparalleled velocities, recognising patterns, irregularities, and potential security menaces.
- **Prompt menace detection:** By means of sophisticated data analytics, potential security hazards can be identified in the present moment. For example, atypical traffic congestion or unexpected variations in environmental parameters can indicate uncommon activities. When these irregularities are recognised, automated notifications can be dispatched to security staff, facilitating prompt reactions and precautionary measures.
- **Forward-thinking reaction tactics:** The fusion of up-to-the-minute data gathering and examination empowers cities to embrace forward-thinking reaction tactics. By recognising burgeoning trends or security apprehensions before they escalate, authorities can allot resources more efficiently. For

instance, if an abrupt surge of individuals is detected in a particular region, extra security staff can be deployed to handle the circumstance.

- **Prognostic analytics:** 5G-enabled IoT systems likewise facilitate the utilisation of prognostic analytics. By scrutinising past data alongside up-to-the-minute inputs, anticipatory models can predict conceivable security occurrences. These models take into account diverse factors, like atmospheric conditions, past criminal trends, and occasion timetables, to foresee security weaknesses and distribute resources accordingly.

- **Augmented emergency services:** In urgent circumstances, up-to-the-minute information plays a pivotal role in maximising emergency reactions. Interconnected IoT devices can offer accurate location information, empowering initial responders to arrive at emergencies more swiftly. Moreover, live video streams from IoT-enabled cameras enable authorities to evaluate situations remotely, augmenting situational consciousness prior to dispatching personnel on-site.

- **Mechanisation and AI-powered replies:** 5G-enabled Internet of Things (IoT) systems open the door for mechanisation and artificial intelligence-powered replies. For example, AI-fueled algorithms can scrutinise live video feeds from surveillance cameras to detect particular occurrences, such as unapproved entry or dubious conduct. These algorithms can activate automated actions, such as dispatching notifications, securing entrances, or redirecting traffic.

In summary, the amalgamation of 5G and IoT technologies empowers intelligent cities with the capability to gather, scrutinise, and react to real-time information with unparalleled velocity and precision. These abilities transform urban security by allowing proactive danger detection, swift response, and the implementation of anticipatory approaches. As cities welcome these progressions, they have the potential to generate more secure and robust surroundings for their inhabitants.

Leveraging IoT Devices for Monitoring and Surveillance

The expansion of Internet of Things (IoT) gadgets has transformed monitoring and surveillance methods, empowering intelligent cities to implement cutting-edge technologies for improved security, situational understanding, and resource maximisation.

- **Extensive data compilation:** IoT devices, furnished with detectors and cameras, provide the capacity to gather immense quantities of data from diverse urban settings. These gadgets can oversee traffic flux, atmospheric condition, sound magnitudes, pedestrian motion, and beyond. The information gathered offers a thorough perspective of urban dynamics, empowering officials to recognise trends and irregularities.

- **Live monitoring:** IoT-enabled surveillance cameras and sensors facilitate live monitoring of public spaces, vital infrastructure, and transportation networks. Ultra-high-definition video feeds can be conveyed in live time to command centres, enabling security personnel to witness occurrences or possible dangers as they transpire.
- **Distant monitoring:** IoT gadgets facilitate distant monitoring, lessening the requirement for physical presence at each location. Law enforcement and security teams have the ability to obtain real-time video streams and sensor information from a distance, enabling them to oversee numerous regions concurrently and react promptly to developing circumstances.
- **Intelligent video analytics:** IoT devices combined with intelligent video analytics leverage artificial intelligence and machine learning to automatically analyze video feeds. These algorithms can identify particular items, actions, or irregularities, such as forsaken parcels or uncommon gathering motions. Automated notifications are produced when pre-established conditions are fulfilled, assisting security staff in recognising possible hazards.
- **Occurrence association and integration:** IoT devices can associate information from various origins, such as surveillance cameras, ecological detectors, and online networking streams. This amalgamation of data provides a more exhaustive comprehension of events. For example, throughout a communal occasion, live information from diverse origins can aid authorities in foreseeing and handling crowd regulation more efficiently.
- **Augmented emergency reaction:** In critical circumstances, IoT devices have a pivotal function in enhancing emergency reaction. Location data from linked devices provides precise details about the occurrence's whereabouts, enabling emergency services to react promptly and allocate resources effectively. Live video streams from IoT cameras assist responders in evaluating the circumstance from a distance.
- **Prognostic analytics for preemptive surveillance:** IoT devices, merged with prophetic analytics, enable cities to foresee security occurrences. By scrutinising past and up-to-the-minute data patterns, authorities can pinpoint potential security menaces before they escalate. This proactive strategy allows effective resource distribution and prompt intervention.
- **Confidentiality considerations:** Although utilising IoT devices for monitoring and surveillance presents various advantages, it also raises confidentiality apprehensions. Finding the perfect equilibrium between security and privacy is vital. Implementing resilient data encryption, ensuring fortified data storage, and establishing explicit guidelines for data usage and retention are imperative to tackle these concerns.

In summary, IoT devices have revolutionised monitoring and surveillance in intelligent cities, offering authorities with instantaneous insights, automated examination, and improved situational consciousness. By leveraging the potential of IoT technology, cities can establish safer and more fortified urban environments while upholding privacy and moral considerations.

Governance is a critical aspect of efficient IIoT integration, particularly in smart manufacturing environments. The authors discuss decision-making authorities, coordination mechanisms, and governance models necessary for optimizing IIoT implementations in smart factories. Understanding these governance principles is essential for harnessing the full potential of IIoT in manufacturing.

DATA PRIVACY AND CITIZEN RIGHTS

In the epoch of 5G and the Internet of Things (IoT) in intelligent metropolises, the utmost significance of data confidentiality and denizen entitlements cannot be overemphasized. The vast information gathering and interconnectivity of IoT devices offer cities with priceless observations for enhanced urban planning and provision of services. Nevertheless, this additionally elevates noteworthy apprehensions regarding the acquisition, retention, and utilisation of individuals' private information. Safeguarding data privacy involves implementing stringent security measures, transparent data practices, and robust encryption protocols to prevent unauthorized access and breaches. Furthermore, as intelligent metropolis endeavours progress, individuals' entitlements to confidentiality, authorization, and dominion over their information must be maintained. Maintaining a fine equilibrium between the advantages of information-based urban administration and upholding individual liberties is crucial to cultivate confidence among inhabitants, guaranteeing that scientific progressions contribute favourably to their existence without jeopardising their fundamental confidentiality and entitlements in the digital era.

Ericsson's "A guide to 5G network security" serves as a comprehensive resource for understanding the security aspects of 5G networks. It covers topics such as network slicing security, authentication mechanisms, and the protection of critical infrastructure. This guide offers valuable insights into the security considerations essential for deploying IoT solutions on 5G networks.

Balancing Data Collection with Citizen Privacy Rights

In the context of the progressing landscape of intelligent cities, attaining a delicate balance between data gathering for urban advancement and protecting citizen privacy rights is a crucial dilemma. The amalgamation of 5G and Internet of Things (IoT) technologies offers unparalleled chances for data-centric decision-making and enhanced urban services, but it also requires watchful endeavours to safeguard individuals' privacy.

- **Clarity and assent:** Harmonising data gathering with privacy privileges involves openly notifying individuals about the information being gathered, its intended purpose, and obtaining their knowledgeable approval. Citizens ought to possess the privilege to comprehend which IoT gadgets are acquiring their data and for what reason, enabling them to formulate knowledgeable choices about their engagement.

- **Reduced data gathering:** Intelligent urban projects ought to concentrate on gathering solely the essential data needed for particular objectives. By reducing the gathering of personally identifiable data and choosing anonymized or consolidated information whenever feasible, cities can find a harmony between accomplishing their goals and safeguarding privacy.

- **Data protection measures:** Enforcing strong data protection measures is crucial. Cryptography, fortified data retention, and stringent entry restrictions aid in protecting gathered information from unauthorised entry or violations. Giving precedence to cyber defence not only safeguards individual confidentiality but also guarantees the soundness of the complete intelligent metropolis framework.

- **Anonymization and de-identification:** Whenever possible, gathered data should be anonymized or de-identified to avoid the association of particular details with individuals. This approach maintains the utility of data for analysis while significantly reducing the risks to citizen privacy.

- **Data life cycle management:** Establishing concise principles for the life cycle of gathered data is vital. Once the information's objective is accomplished, it should be erased or made anonymous promptly. Appropriate data administration practices avert the superfluous preservation of individual data and diminish potential confidentiality hazards.

- **Empowering citizen authority:** Citizens ought to possess the capability to obtain, examine, and govern the information gathered concerning them. Offering individuals with mechanisms to assess and modify their data guarantees that they possess control over their personal information and can correct any errors.

- **Supervision and control:** Regulatory structures and supervision mechanisms play a crucial role in balancing data gathering and confidentiality privileges. Authorities and governing bodies have the ability to set forth principles, norms, and sanctions for failure to comply, guaranteeing that intelligent urban projects adhere to moral data protocols (Walia *et al.*, 2019).

In summary, attaining equilibrium between data gathering for intelligent urban development and honouring citizen privacy entitlements necessitates a multifaceted strategy. Lucid communication, reduced data gathering, rigorous security measures, and empowering individuals with authority over their data collectively contribute to establishing an intelligent urban environment that upholds confidentiality while reaping the advantages of technological progress.

Legal and Ethical Considerations in Collecting and Using Citizen Data

In the ever-changing terrain of intelligent metropolises fueled by 5G and Internet of Things (IoT) technologies, the gathering and application of public data introduce a plethora of lawful and moral considerations that must be prudently maneuvered to guarantee the conscientious and considerate utilisation of data.

- **Data possession and approval:** Intelligent urban projects must tackle inquiries of data possession. Citizens should possess transparency regarding who possesses the data gathered from them and the manner in which it will be utilised. Acquiring knowledgeable and unequivocal consent from individuals prior to gathering their data is a fundamental ethical prerequisite, empowering citizens to govern how their information is employed.
- **Conformity with regulations:** Abiding by data protection regulations is crucial. Intelligent urban initiatives necessitate harmonisation with local and nationwide legislations, like the Comprehensive Data Security Directive (CDSD) in Europe. Conformity guarantees that citizen information is managed conscientiously and that individuals' entitlements to confidentiality are maintained.
- **Objective restriction:** Gathering information for distinct, valid objectives and abstaining from utilising it for unrelated undertakings honours citizens' anticipations and entitlements. Information should be gathered solely for clearly defined goals and not reused without explicit agreement.
- **Data reduction:** Restricting the gathering of personally identifiable data to what is essential for the intended objective lessens privacy hazards. Accumulating surplus data amplifies the likelihood of abuse and infringements, rendering data reduction an ethical principle to adhere to.
- **Anonymization and de-identification:** Masking or depersonalising gathered data aids in safeguarding individuals' confidentiality by inhibiting the association of data with particular persons. Morally, this procedure guarantees that data examination and perceptions can happen without jeopardising individual identities.
- **Clarity and correspondence:** Candidly conveying data gathering methodologies, objectives, and possible hazards nurtures confidence between citizens and intelligent urban projects. Translucency guarantees that individuals comprehend how their information is being utilised and enables them to make knowledgeable choices.
- **Data safeguarding and preservation:** Guaranteeing the safety of gathered data is a moral obligation. Implementing robust cybersecurity measures protects citizen information from unauthorised access, breaches, and potential abuse.
- **Justice and equality:** Reflections on justice and equality should steer data-gathering practices to prevent unduly impacting specific demographics. Moral considerations encompass alleviating prejudices that may emerge from partial

data or algorithms, guaranteeing that intelligent urban area advantages are reachable to all inhabitants.

- **Public responsibility:** Smart city initiatives should be responsible to the public. Creating systems for individuals to voice their worries, obtain their information, and pursue compensation in the event of violations or abuse is crucial to upholding confidence.

- **Continuous assessment and improvement:** Ethical considerations should drive ongoing assessment and improvement of data collection practices. Consistently examining data protocols and tackling emerging confidentiality worries aids in upholding moral principles in a swiftly developing environment (Kim *et al.*, 2019).

In conclusion, the collection and use of citizen data in smart cities demand adherence to legal regulations and ethical principles. Achieving a harmonious equilibrium between novelty and regard for citizen entitlements necessitates meticulous contemplation of data possession, openness, safeguarding, impartiality, and answerability. By maintaining these considerations, intelligent cities can guarantee that technological progress is in harmony with moral principles and contribute favourably to the lives of residents.

Faheem *et al.* (2018) explore the intersection of smart grid communication and Industry 4.0. This review discusses the role of 5G and IoT technologies in modernizing the energy sector, emphasizing their potential in enhancing grid reliability, optimizing energy distribution, and enabling demand response. Understanding these dynamics is crucial for comprehending the synergy between 5G, IoT, and smart grid applications.

Implementing Transparent Data Practices to Build Citizen Trust

In the backdrop of 5G-facilitated intelligent urban areas, constructing and upholding citizen confidence is of utmost importance. Lucid data practices play a crucial role in nurturing confidence by guaranteeing that individuals comprehend how their data is gathered, utilised, and safeguarded. Here is how translucent data practices can be implemented to establish and nurture citizen confidence:

1. Transparent Data Guidelines and Assent.

2. Lucid Data Regulations and Approval.

3. Obvious Data Protocols and Agreement.

4. Evident Data Measures and Concurrence.

5. Apparent Data Principles and Accord.

6. Plain Data Rules and Compliance.

7. Explicit Data Procedures and Permission.

Formulate extensive data guidelines that delineate the objective of data acquisition, varieties of data gathered, and the manner in which it shall be utilised. Offer transparent and succinct elucidations to individuals regarding their entitlements and the means to grant knowledgeable agreement. Clarity guarantees individuals are completely cognizant of what they are consenting to when disclosing their information.

- **Simple-to-comprehend communication:** Utilise straightforward language to convey data practises, evading intricate terminology that might perplex or isolate citizens. Knowledge should be readily obtainable *via* websites, mobile applications, and alternative communication channels, empowering individuals to form enlightened choices regarding their information.
- **Data utilisation clarity:** Elaborate on the precise methods through which gathered data will be employed to enrich services, enhance infrastructure, or render urban life more effective. Citizens ought to comprehend the immediate advantages that arise from information exchange, cultivating a feeling of collaboration between them and the intelligent urban project.
- **Consent administration:** Integrate user-centric consent administration systems that enable individuals to regulate their choices for data gathering and utilisation. Individuals should possess the capacity to conveniently alter or retract their agreement at any given moment, granting them authority over their information.
- **Data accumulation alerts:** Inform individuals when their information is being gathered *via* IoT devices. Educate them about the objective, the category of information being gathered, and how it will contribute to the enhancement of the city. Lucid notifications empower citizens to make enlightened decisions about participation.
- **Choose-in and choose-out mechanisms:** Offer transparent opt-in and opt-out mechanisms for data gathering. Inhabitants should possess the option to engage willingly and retract their approval if they alter their perspectives. Honouring individuals' independence constructs confidence by recognising their inclinations.
- **Information retention and safeguarding:** Openly convey how gathered data will be stored and safeguarded. Emphasise the precautions implemented to safeguard data from unauthorised entry, infringements, or abuse. Reassuring individuals of strong security measures boosts their trust in data management.
- **Consistent upgrades and documentation:** Consistently upgrade citizens on the utilisation of their data and the results accomplished through intelligent urban

centre endeavours. Sharing triumph tales, enhancements, and revelations showcases the palpable advantages of data sharing, fortifying reliance.

- **Feedback pathways:** Establish avenues for citizens to provide feedback, inquire, and express apprehensions about data practises. Adaptable involvement illustrates a dedication to transparency and constructs confidence by tackling public feedback.
- **Autonomous audits and supervision:** Ponder engaging external auditors or supervision entities to autonomously evaluate data practises. External verification contributes to the credibility and strengthens the dedication to open and moral data management.

In summary, see-through data practises are crucial in constructing and maintaining citizen confidence in intelligent urban projects. By transparently communicating data aggregation, utilisation, and safeguarding measures, cities exhibit a dedication to conscientious data administration that corresponds with citizens' principles and anticipations. Lucid practises guarantee that the advantages of technological progress are accompanied by a robust groundwork of faith and responsibility.

CYBERSECURITY MEASURES IN SMART CITIES

Cybersecurity precautions in intelligent metropolises are crucial to protect interconnected systems and delicate information against advancing cyber hazards. Resilient strategies encompass diverse approaches, including network partitioning, intrusion detection systems, periodic security evaluations, and staff education. Cryptography and verification protocols guarantee information secrecy and hinder unauthorised entry. Occurrence reaction blueprints empower prompt actions in the occurrence of infringements, while ongoing surveillance and menace understanding sharing fortify preparedness. Cooperative endeavours amidst public and private sectors foster knowledge interchange and collective safeguarding. Ultimately, efficient cybersecurity measures alleviate hazards, guarantee the soundness of vital infrastructure, and uphold citizen confidence, bolstering the sustainable expansion of safe and durable intelligent cities.

Establishing Secure Communication Protocols for IoT Devices

In the elaborate ecosystem of intelligent cities, the formation of secure communication protocols for Internet of Things (IoT) devices is of utmost significance to protect data authenticity, user confidentiality, and overall system dependability. Protected communication protocols serve as the groundwork for ensuring that data transmitted between IoT devices and central servers remains safeguarded against diverse cyber threats.

Cryptography for information privacy: Reliable communication protocols utilise encryption methods to guarantee that data is conveyed in an indecipherable format, efficiently protecting it from unauthorised interception and eavesdropping. This hinders delicate data from being jeopardised during transmission, even if malevolent performers acquire entry to the network.

Verification methods: Robust authentication mechanisms are incorporated into secure protocols to authenticate the identities of both the transmitting and receiving parties. This hinders unapproved devices from acquiring entry to the network and guarantees that information is exchanged solely between validated and sanctioned devices.

Data authenticity guarantee: Protected protocols frequently incorporate mechanisms for ensuring data integrity. This implies that the information being conveyed remains unchanged during transportation. Hash algorithms and electronic signatures are frequently employed to authenticate that the received information has not been altered while being transmitted.

Reciprocal authentication: In further sophisticated configurations, reciprocal authentication is utilised, necessitating both the IoT device and the server to authenticate one another. This dual-factor verification improves safety by obstructing unauthorised devices from connecting to the network and safeguarding against deceitful servers that try to intercept data.

Over-the-air updates with safety: Wireless (OTA) updates are crucial for maintaining IoT devices up to date with the most recent security fixes. Protected communication protocols guarantee that updates are verified and ciphered, hindering unauthorised parties from infusing malevolent updates into the devices.

Uniformity and accreditation: Standardising safeguarded communication protocols across IoT devices and platforms is crucial. Validation programmes guarantee that devices conform to security criteria prior to being implemented. These endeavours foster uniformity and enhance the overall safety stance of intelligent urban environments.

Obstacles and reflections: Implementing fortified communication protocols necessitates meticulous deliberation of elements such as delay, energy expenditure, and computational burden. Achieving a harmonious equilibrium between sturdy security and pragmatic usability is crucial to prevent impeding the functionality of IoT devices.

In summary, establishing fortified communication protocols for IoT devices is a foundation of intelligent urban cybersecurity. These measures guarantee the secrecy, soundness, and genuineness of information exchanged, safeguarding individuals' confidentiality and vital infrastructure from cyber hazards. As the cornerstone of secure data transmission, these protocols play a crucial role in establishing a resilient and reliable foundation for the intelligent cities of the future.

Intrusion Detection and Prevention Systems for Critical Infrastructure

The execution of intrusion detection and prevention systems (IDPS) is vital for protecting the authenticity and safety of essential infrastructure in intelligent cities. These systems are crafted to preemptively detect and hinder unauthorised entry, malevolent actions, and cyber assaults that aim at vital systems and services.

Perpetual surveillance: Intrusion Detection and Prevention Systems persistently oversee network traffic, system conduct, and data configurations to identify any irregularities or divergences from typical operations. This live monitoring facilitates swift recognition of potential security violations.

Anomaly detection: IDPS utilise advanced algorithms to scrutinise network traffic and system conduct, contrasting it with established benchmarks. Divergences from these benchmarks activate notifications, suggesting possible infiltrations or security violations.

Signature-based identification: IDPS utilise signature-based identification techniques that compare incoming data against a repository of recognised assault patterns or signatures. If a match is discovered, the system can take prompt action to avert the assault.

Behavioural examination: Behavioural scrutiny methods detect atypical trends of conduct that might not correspond with recognised assault indications. This technique is efficient in identifying formerly unfamiliar dangers or zero-day susceptibilities.

Alert creation: When a breach endeavour or malevolent action is identified, IDPS produce notifications that are dispatched to security personnel or administrators. These notifications encompass details regarding the essence of the peril, its origin, and suggested measures for alleviation.

Precautionary measures: Furthermore, IDPS can be set up to implement precautionary measures, such as obstructing or segregating the origin of the danger. This proactive strategy aids in minimising potential harm and lessening the influence of cyber assaults.

Incorporation with security operations: Intrusion Detection and Prevention Systems (IDPS) are frequently integrated with wider security operations centres (SOCs), enabling centralised administration, examination of menace information, and synchronisation of reaction endeavours.

Scalability and flexibility: IDPS can be expanded to accommodate the developing and enlarging nature of critical infrastructure. They can adjust to novel assault vectors and emerging menaces, guaranteeing a proactive safeguarding mechanism.

Obstacles and reflections: Efficient execution of IDPS necessitates a comprehensive comprehension of the infrastructure's distinct attributes, possible susceptibilities, and the capacity to distinguish authentic irregularities from malevolent actions. Inaccurate positives and inaccurate negatives should be reduced to prevent disturbance and guarantee effective response.

In summary, intrusion detection and prevention systems are essential elements of cybersecurity strategies for safeguarding vital infrastructure in intelligent cities. By expeditiously recognising and addressing potential hazards, these systems aid in guaranteeing the uninterrupted functioning, durability, and safeguarding of vital services, rendering intelligent cities more sturdy and less susceptible to cyber intrusions.

GSMA's report on "5G Use Cases for Verticals" outlines various industry-specific applications of 5G technology. It discusses the potential impact of 5G and IoT on sectors such as healthcare, agriculture, and manufacturing. Understanding these diverse use cases is essential for grasping the breadth of opportunities offered by 5G and IoT integration across industries.

Encryption and Authentication Techniques for Safeguarding Data

In the interconnected terrain of intelligent metropolises, the utilisation of encryption and validation methods is crucial to protect delicate information from unauthorised entry, interception, and alteration. These methods offer strata of security that safeguard data integrity, privacy, and genuineness.

• **Encryption for Data Confidentiality:** Encryption metamorphoses data into an indecipherable format using cryptic algorithms. Only authorised entities with the decryption code can decode the data, guaranteeing that even if intercepted during transmission, the information remains protected. Complete encryption safeguards information from the transmitter to the recipient, obstructing snooping and unapproved entry.

- **Public-Key Infrastructure (PKI):** PKI encompasses the utilisation of communal and confidential key pairs for encryption and verification. Public keys are extensively dispersed and employed to encode information, whereas private keys are safeguarded and utilised for deciphering. This guarantees safe communication between parties, as exclusively the intended receiver possesses the confidential key.
- **Verification Mechanisms:** Verification validates the identity of users, devices, or systems prior to granting entry to confidential information. Multi-faceted authentication (MFA) necessitates various types of validation, like passcodes, biometric data, or tokens, rendering it more challenging for unauthorised individuals to acquire entry.
- **Two-Factor Verification (2FV) and Multi-Factor Verification (MFV):** 2FV and MFV incorporate an additional stratum of protection by necessitating users to furnish various types of identification prior to gaining entry to data or systems. This lessens the peril of unauthorised entry even if one element, such as a passphrase, is jeopardised.
- **Tokenization:** Tokenization entails substituting confidential information with distinct tokens that possess no intrinsic worth. Even if intercepted, tokens are futile to assailants. Tokenization is frequently employed for safeguarding payment card details and other confidential data.
- **Certificate-Based Authentication:** Certificates are electronic testimonials that validate the identity of devices, users, or systems. They are distributed by reliable certificate authorities (CAs) and guarantee secure communication between parties by verifying identities.
- **Obstacles and Reflections:** While encryption and verification methods greatly improve data safety, obstacles involve key administration, expandability, and guaranteeing a smooth user encounter. Finding an equilibrium between sturdy security and user-friendliness is pivotal to avert obstacles to authorised entry.

In summary, encryption and verification methods are essential to safeguarding confidential information in intelligent urban areas. These measures ensure data privacy, deter unauthorised entry, and ensure the genuineness of parties involved in communication. By utilising these methods, intelligent cities can establish a secure and reliable digital atmosphere that protects citizen confidentiality and vital data.

CASE STUDIES: SECURE SMART CITY IMPLEMENTATIONS

Analysing tangible instances of fortified intelligent urban developments offers valuable perspectives into the efficient amalgamation of 5G, IoT, and information security measures.

Singapore's Smart Nation Initiative: Prioritizing Cybersecurity

- **Context and aims:** The Intelligent Nation Campaign is Singapore's tactical endeavour to utilise technology and ingenuity to establish a flawlessly linked metropolis where technology enhances the standard of living for its inhabitants. The proposal envisions a clever metropolitan ecosystem where information and

  technology amplify different facets of everyday life, from mobility and wellness to city design and civic amenities.
- **Cybersecurity emphasis:** Acknowledging the vast capacity of technology also accompanies heightened cybersecurity hazards, Singapore places a robust focus on cybersecurity within its Intelligent Nation Campaign. This forward-thinking approach recognises that as digital revolution quickens, ensuring the foundational digital framework becomes crucial to protect confidential data, vital services, and public confidence.
- **Data security and cryptography methods:** In line with its cybersecurity plan, Singapore's Smart Nation Initiative utilises sophisticated encryption methods to safeguard citizen data. Encryption entails converting information into a safeguarded, indecipherable structure employing intricate formulas. Sophisticated encryption guarantees that even if unauthorised parties acquire access to the data, they cannot unravel it without the suitable decryption key.
- **Cooperative approach:** Singapore's strategy to cybersecurity within the Smart Nation Initiative is distinguished by cooperation among diverse stakeholders. Government entities, educational institutions, business collaborators, and cyber defence specialists collaborate to formulate and execute all-encompassing safeguarding protocols. This cooperative endeavour guarantees a comprehensive and all-encompassing approach to tackling cybersecurity obstacles, as diverse viewpoints contribute to recognising and alleviating hazards efficiently.
- **Holistic security stance:** Through promoting cooperation among these varied entities, Singapore's Intelligent Nation Initiative accomplishes an all-encompassing security stance. This implies that cybersecurity considerations are ingrained at each phase of the digital metamorphosis process, from conceptualising and executing intelligent urban solutions to continuous surveillance and reaction.
- **Invention and safety:** Significantly, Singapore's Intelligent Country Initiative illustrates that giving importance to cyber defence does not impede novelty; instead, it amplifies it. A safe and durable digital framework nurtures a setting where individuals, enterprises, and administrations can confidently adopt novel technologies and offerings, propelling economic expansion and enhancing quality of existence.
- **Public confidence and involvement:** The focus on cyber defence not only guarantees the safety of citizens' information but also nurtures public reliance in

the endeavours of the Intelligent Society. Citizens are more inclined to engage in digital services and disclose data when they are guaranteed that their privacy and information are adequately safeguarded.

In summary, Singapore's Intelligent Nation Initiative showcases how a progressive city can prioritise cyber defence to construct a secure and robust digital framework. By concentrating on data safeguarding, utilising encryption methods, and nurturing cooperative endeavours, Singapore establishes a milieu where ingenuity flourishes, public confidence is upheld, and the potential advantages of a linked metropolis are achieved while mitigating the accompanying hazards.

Barcelona's CityOS Platform: Enhancing Data Privacy and Trust

- **Platform Overview:** The CityOS framework in Barcelona is a paragon of safeguarded data methodologies within the context of an intelligent metropolis. It functions as a centralised framework that gathers and evaluates up-to-the minute information from various origins, encompassing Internet of Things (IoT) gadgets deployed throughout the urban area. This information-based approach seeks to enhance urban services, boost effectiveness, and establish a more habitable environment for inhabitants.
- **Data Confidentiality Emphasis:** In the age of interconnected devices, guaranteeing data confidentiality is vital. Barcelona's CityOS platform puts a notable focus on protecting citizens' information and their entitlement to confidentiality. By engaging in such actions, the urban area showcases its dedication to conscientious data administration and upholding the confidence of its inhabitants.
- **Cryptography and Entry Restrictions:** In order to safeguard the information gathered by the CityOS platform, encryption methods are utilised. Cryptography converts information into an indecipherable structure, and solely authorised individuals with decryption codes can obtain and decode it. Furthermore, access restrictions are enforced to restrict who can observe and handle the data, guaranteeing that confidential information remains safeguarded.
- **Finding the Equilibrium:** A primary obstacle for intelligent metropolises is to discover an equilibrium between information-fueled urban administration and individual confidentiality. Barcelona's CityOS platform manoeuvres this dilemma by incorporating rigorous security measures that protect data while enabling the city to utilise the knowledge acquired from the data for enhanced decision-making.
- **Amplifying Confidence and Engagement:** By prioritising data confidentiality and implementing resilient security measures, Barcelona nurtures a feeling of reliance among its citizens. When inhabitants trust that their information is

managed conscientiously, they are more inclined to engage in municipal endeavours that entail exchanging data. This involvement is crucial for the triumph of intelligent urban projects, as it offers valuable perspectives that can propel favourable transformation.

- **Openness and Responsibility:** A significant facet of the CityOS platform is its openness in data practises. By conveying to citizens how their information is gathered, utilised, and safeguarded, Barcelona guarantees that residents are well-instructed participants in the city's data environment. This translucency improves responsibility and strengthens the municipality's dedication to moral information administration.

Final thought: Barcelona's CityOS platform exhibits how safe data practises can be incorporated into an intelligent city setting. By gathering and scrutinising information conscientiously, utilising encryption and entry restrictions, and emphasising openness, Barcelona not only showcases its dedication to individual confidentiality but also establishes a milieu where inhabitants are more inclined to get involved and partake in endeavours that utilise data for the enhancement of city existence.

Songdo, South Korea: Secure Communication Through Dedicated Networks

- **Intelligent City Infrastructure:** Songdo, situated in South Korea, is celebrated for its inventive strategy to metropolitan planning and technology incorporation. As an intelligent metropolis, Songdo utilises cutting-edge technologies to amplify diverse facets of urban existence, ranging from mobility and energy optimisation to healthcare and civic involvement.
- **Significance of Committed Networks:** In the quest for establishing a safe and interconnected intelligent metropolis, Songdo emphasises the significance of specialised networks for Internet of Things (IoT) gadgets. Internet of Things (IoT) gadgets, like detectors and linked devices, have a vital function in gathering and transmitting information that enlightens diverse urban operations. Nevertheless, the interlinked nature of these gadgets also exposes them to possible cyber hazards.
- **Committed Networks for IoT Traffic:** To tackle this susceptibility, Songdo establishes distinct and dedicated networks specifically crafted for IoT traffic. These networks are segregated from other communication channels utilised for vital services and overall internet connectivity. By partitioning IoT traffic, the city reduces the chance of unauthorised entry and cyber assaults aimed at sensitive information and vital infrastructure.
- **Improved Security:** Devoted networks provide improved security by decreasing the vulnerability area. Given that Internet of Things (IoT) devices are segregated from alternative networks, potential assailants have restricted

pathways for penetrating the urban area's systems. This segregation functions as a barricade, rendering it notably more challenging for cyber dangers to spread from IoT devices to central infrastructure.

- **Diminished Assault Path:** The segregation of IoT communication into specialised networks lessens the assault path for malevolent individuals endeavouring to capitalise on weaknesses in the IoT framework. Even if a violation transpires within the IoT network, the repercussions are confined within that segregated milieu, hindering sideward progression to other segments of the intelligent urban framework.

- **Preventing Unsanctioned Entry:** Through establishing dedicated networks, Songdo guarantees that solely authorised devices can communicate within the IoT network. Unsanctioned devices are efficiently hindered from accessing or disrupting the municipality's vital systems, preserving the integrity and functionality of necessary services.

- **Boosting Resilience:** The partitioning of IoT traffic into specialised networks amplifies the overall robustness of Songdo's intelligent urban infrastructure. In the occurrence of a cyber assault aiming at the IoT ecosystem, the consequence is confined, enabling the city to react more efficiently and promptly to confine and alleviate the menace.

- **Conclusion:** The illustration of Songdo, South Korea, emphasises the significance of specialised networks for IoT devices in upholding secure communication within an intelligent urban environment. By segregating IoT traffic and establishing exclusive channels, Songdo exemplifies a forward-thinking stance towards cybersecurity, diminishing the peril of cyber assaults and guaranteeing the soundness of interconnected systems. This strategy corresponds with the municipality's dedication to constructing a resilient and safeguarded urban setting for its inhabitants.

These case studies highlight the importance of preemptive cybersecurity measures in intelligent urban deployments discussed in Table **1**. They emphasise the incorporation of reliable communication protocols, information confidentiality measures, and cooperative endeavours among participants to establish resilient and safeguarded intelligent urban landscapes. By acquiring knowledge from these prosperous instances, cities globally can embrace analogous approaches to construct resilient and impregnable urban ecosystems that exploit the advantages of technology while protecting citizen data and vital infrastructure discussed in Table **2**.

Table 1. Case studies on secure smart city implementations [10].

City	Approach to Cybersecurity	Key Security Measures
Singapore's Smart	Emphasis on cybersecurity within the Intelligent Nation Campaign	- Use of advanced encryption methods to safeguard citizen data - Cooperative approach among diverse stakeholders - Ingrained cybersecurity considerations at each phase of the digital transformation process
Nation Initiative	Utilization of encryption methods for data security	- Promotion of innovation and safety - Enhancement of public confidence and involvement
Barcelona's CityOS	Focus on data confidentiality and individual privacy	- Use of encryption and access restrictions - Finding an equilibrium between data-driven urban administration and individual privacy - Amplifying confidence and engagement among citizens
Platform	Transparency in data practices	- Openness and responsibility in conveying data collection, usage, and protection practices
Songdo, South Korea's	Use of dedicated networks for IoT traffic Improved security	- Enhanced security by reducing the vulnerability area - Diminished assault path for cyber threats - Prevention of unsanctioned entry - Boosting resilience in the event of a cyberattack
Intelligent City	Prevention of unauthorized access	-
Infrastructure	Enhancement of overall system resilience	-

Table 2. Key security measures in secure smart city implementations [11].

Security Measure	Description
Emphasis on cybersecurity	Prioritizing cybersecurity within smart city initiatives, recognizing the increased cyber threats in the digital era.
Advanced encryption methods	Utilizing sophisticated encryption techniques to protect citizen data, making it indecipherable to unauthorized parties.
Cooperative approach	Collaborating among various stakeholders, including government entities, educational institutions, businesses, and cybersecurity specialists, to formulate and execute comprehensive security protocols.

(Table 2) cont.....

Security Measure	Description
Ingrained cybersecurity considerations	Ensuring that cybersecurity considerations are integrated into every phase of the digital transformation process, from conceptualization to continuous monitoring and response.
Promotion of innovation and safety	Demonstrating that a focus on cybersecurity does not hinder innovation but rather enhances it by creating a secure environment for technology adoption.
Enhancement of public confidence	Building trust among citizens by assuring the safety of their data, leading to increased participation in digital services and data sharing.
Data confidentiality and privacy emphasis	Placing notable emphasis on protecting citizens' data and their right to privacy in an interconnected urban environment.
Transparency in data practices	Being transparent with citizens about how data is collected, used, and protected, fostering responsible data management and ethical data administration.
Dedicated networks for iot traffic	Creating distinct and dedicated networks specifically designed for IoT traffic to enhance security and isolate IoT devices from other critical infrastructure.
Prevention of unauthorized access	Ensuring that only authorized devices can communicate within the IoT network, preventing unauthorized devices from accessing or disrupting vital systems.
Overall system resilience enhancement	Improving the resilience of the entire urban infrastructure by containing the impact of cyberattacks on the IoT ecosystem, enabling more efficient responses to threats.

Examining Successful Examples of Smart Cities with Robust Security Measures

- **Singapore, Singapore:** Singapore's Intelligent Nation Initiative is a prime exemplification of a metropolis that prioritises cyber defence. The initiative concentrates on data security, utilises cutting-edge encryption methods to protect citizen data, and highlights cooperative endeavours between government organisations, academia, and industry collaborators. The municipality's extensive strategy to safeguarding guarantees that novelty and technological progressions go hand in hand with upholding a robust security stance.

- **Copenhagen, Denmark:** Copenhagen's strategy for intelligent urban security encompasses transparent data principles combined with rigorous privacy safeguarding. The metropolis highlights openness and grants residents authority over their information *via* its "Data Morality" campaign. Copenhagen concentrates on ensuring the gathering, retention, and exchange of data while enabling citizens to reach and govern their facts, thereby amplifying reliance and engagement.

- **Dubai, UAE:** Dubai's Intelligent Dubai initiative incorporates cutting-edge technologies while upholding a robust cybersecurity framework. The metropolis utilises cutting-edge cybersecurity solutions such as synthetic intelligence (AI)-

enabled menace detection, instantaneous surveillance, and flexible reaction mechanisms. Dubai's devotion to cybersecurity is apparent through its devoted Cyber Resilience Strategy and partnerships with worldwide cybersecurity specialists.

- **Taipei, Taiwan:** Taipei's "Intelligent Taipei" campaign highlights information security and confidentiality by embracing blockchain technology for data authentication and exchange. The metropolis utilises blockchain to guarantee openness and traceability in data transactions, augmenting data integrity and lessening the peril of unauthorised entry. Taipei's emphasis on data security is in sync with its objective of establishing a safe and citizen-oriented intelligent metropolis atmosphere.

- **Barcelona, Spain:** Barcelona's CityOS platform exhibits reliable data practises through encryption and entry controls. By giving precedence to data confidentiality and achieving an equilibrium between data-focused urban administration and citizen confidentiality, Barcelona improves reliance and involvement in its undertakings. The metropolis's strategy emphasises the significance of safeguarding information while utilising its capacity for enhanced urban amenities.

- **Songdo, Republic of Korea:** Songdo exemplifies the significance of specialised networks for IoT devices in guaranteeing secure communication. By establishing distinct networks for IoT traffic, the city prevents unauthorised entry to vital infrastructure and delicate information. This segregation amplifies safety and diminishes the peril of cyber assaults targeting interconnected systems.

These prosperous instances demonstrate the varied tactics that intelligent cities utilise to enforce resilient security measures. By giving precedence to cyber safety, utilising encryption and verification methods, and promoting cooperation among parties involved, these cities establish atmospheres where ingenuity and technology coexist with robust data safeguarding and public confidence.

COLLABORATIVE EFFORTS FOR SECURITY ENHANCEMENT

Collective endeavours are crucial for augmenting security in intelligent urban areas, as the intricacy of cyber defence predicaments necessitates a consolidated strategy. Public-private collaborations, knowledge exchange among cities, and cooperation between cybersecurity specialists and government entities are fundamental elements of these endeavours shown in Fig. (**5**). By combining assets, knowledge, and menace intelligence, cities can collectively tackle emerging cyber dangers, cultivate optimal approaches, and execute proactive safeguard measures that establish a sturdier and more resilient groundwork for secure urban settings.

Role of Government, Industry, and Academia in Building Secure Smart Cities

- Administration, sector, and educational institutions each have separate yet interconnected functions in collectively constructing safe intelligent metropolises that prioritise citizen security and information safeguarding.
- **Administration:** Administrations play a crucial role in establishing the regulatory structure and benchmarks for security in intelligent cities. They establish cyber defence policies, data safeguard regulations, and adherence prerequisites that direct the implementation of security measures. Government authorities additionally offer the essential supervision and synchronisation to guarantee that security measures are uniform across various intelligent urban initiatives. By means of financing, motivations, and collaborations between the public and

private sectors, authorities promote creativity and stimulate the collaboration between businesses and educational institutions to offer their knowledge.

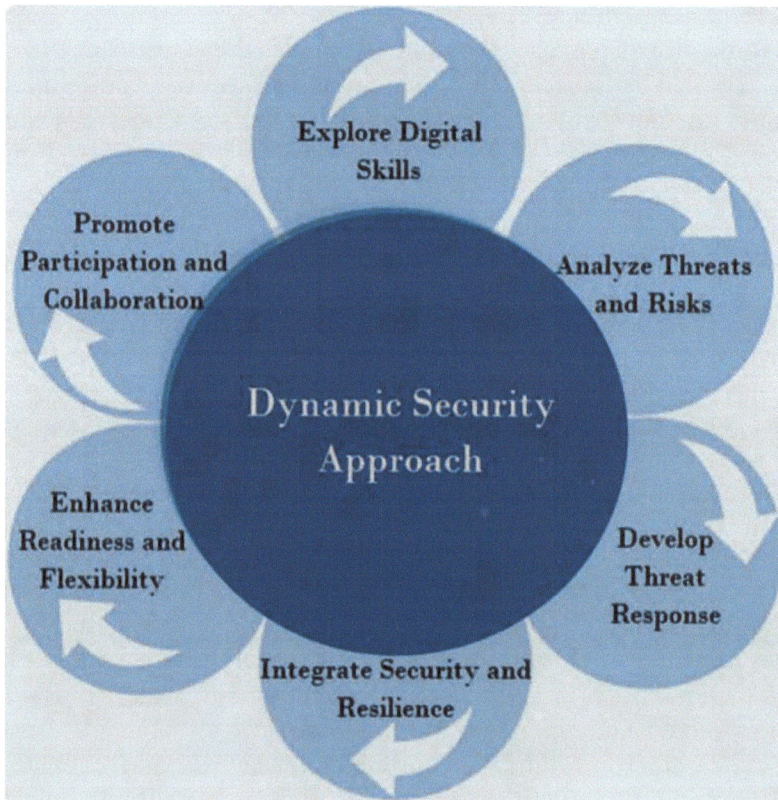

Fig. (5). Dynamic security approach [15].

- **Sector:** The technology sector propels the advancement and execution of intelligent urban solutions, encompassing the creation of fortified frameworks, gadgets, and software. Business partners develop and incorporate cutting-edge security technologies such as ciphering, verification, intrusion identification systems, and safeguarded communication protocols. They furnish the instruments and resolutions necessary to protect data, administer entry commands, and react efficiently to cyber menaces. Industry cooperation with government agencies ensures that security prerequisites are fulfilled while delivering inventive solutions that enrich urban existence.

- **Scholasticism:** Scholarly establishments contribute exploration, proficiency, and adept specialists to the advancement of fortified intelligent urban areas. They perform investigations on emerging cyber menaces, weaknesses, and safeguarding resolutions. Scholarship plays a crucial function in educating and instructing the labour force required to execute, oversee, and enhance safety precautions in intelligent urban settings. Working together with government and industry collaborators, academia nurtures an atmosphere of ongoing education and information interchange that propels the development of efficient security methodologies. Synergy and Cooperation: The collaboration among government, industry, and academia is vital for constructing secure intelligent cities. Authorities furnish the regulatory structure, sector delivers inventive technologies, and academia presents research-based perspectives. Cooperative endeavours enable the amalgamation of optimal methodologies, state-of-the-art resolutions, and all-encompassing safety approaches that tackle developing menaces. Collectively, these stakeholders collaborate to establish urban settings that harness technology for the welfare of residents while guaranteeing data confidentiality, cyber protection, and citizen well-being are of utmost importance.

Creating Partnerships to Share Best Practices, Research, and Resources

Forging alliances among diverse stakeholders is crucial for intelligent cities to efficiently tackle security obstacles and construct robust digital ecosystems. These collaborations facilitate the exchange of optimal methodologies, exploration discoveries, and assets, ultimately contributing to improved cyber protection and the triumph of intelligent urban projects.

- **Inter-Sector Cooperation:** Cooperative alliances among governmental organisations, business pioneers, educational establishments, and cybersecurity specialists promote the interchange of perspectives and proficiency from varied sectors. Inter-sector cooperation guarantees a holistic approach to security, leveraging the advantages of every participant to jointly tackle intricate security obstacles.

- **Sharing Optimal Methods:** Through the act of sharing optimal methods, cities can acquire knowledge from triumphant security strategies and incorporate verified approaches in their own intelligent urban initiatives. Insights acquired from prior executions aid in evading typical traps and offer direction for constructing resilient security measures.
- **Research Collaboration:** Educational establishments provide state-of-the-art investigation on cyber defence, menace awareness, and burgeoning innovations. Cooperating with researchers enables cities to harness the most recent perspectives to remain ahead of cyber risks and adjust security measures to developing challenges.
- **Resource Aggregation:** Collaborations facilitate the aggregation of resources, both monetary and technological. Collective endeavours harness pooled resources to invest in cutting-edge security technologies, carry out thorough security evaluations, and establish ongoing surveillance systems.
- **Wisdom Swap:** Sharing wisdom *via* workshops, conventions, and instructional sessions nurtures a culture of perpetual education. This wisdom interchange amplifies the proficiency of security experts, empowering them to execute and oversee security precautions efficiently.
- **Global Networks:** International alliances and networks link intelligent cities globally, enabling the exchange of experiences and remedies on a worldwide level. Acquiring knowledge from various regions' security practices aids cities in comprehending varied threat landscapes and adjusting applicable strategies.
- **Public-Private Alliances:** Cooperation amidst public and private sectors utilises the capabilities of both entities. Governments provide regulatory counsel, while industry partners offer inventive technologies and resolutions. This collaboration leads to the execution of safety measures that correspond with industry top practises and regulatory obligations.
- **Creativity and Flexibility:** Collaborations promote ingenuity by motivating the exploration of novel methods to security obstacles. Collaboration enables cities to adjust to evolving threat scenarios and cultivate inventive solutions that foresee forthcoming hazards.

In summary, collaborations that enable the exchange of optimal methodologies, exploration, and assets are essential to constructing safe intelligent urban areas. By utilising the combined knowledge of government, industry, academia, and cybersecurity experts, cities can cultivate strong security measures that safeguard vital infrastructure, data confidentiality, and citizen welfare, ultimately contributing to the progression of secure and durable urban settings.

Establishing Regulatory Frameworks to Ensure Standardized Security Practices

Establishing regulatory structures is crucial for intelligent cities to guarantee standardised security measures that safeguard citizen information, vital infrastructure, and general urban robustness. These frameworks offer principles, prerequisites, and conformity measures that establish the groundwork for resilient cybersecurity in intelligent urban projects.

- **Consistent Guidelines:** Regulatory structures establish consistent guidelines and optimal approaches for safety throughout various intelligent urban initiatives. They guarantee uniformity in security implementations, lessening susceptibilities that can emerge from diverse approaches.
- **Data Security:** Legislative directives tackle data security and confidentiality worries, delineating how individual information ought to be gathered, retained, manipulated, and exchanged. Criteria such as encoding and information deidentification safeguard delicate data while facilitating data-centric urban administration.
- **Cybersecurity Measures:** Regulatory frameworks stipulate cybersecurity measures that must be executed, such as encoding, verification protocols, intrusion detection systems, and incident response strategies. These measures augment the city's capacity to avert, discern, and alleviate cyber hazards.
- **Conformity Obligations:** Urban areas, merchants, and service providers must abide by conformity obligations specified in regulatory frameworks. Conformity guarantees that safety precautions are regularly implemented and that stakeholders are responsible for upholding a protected atmosphere.
- **Risk Evaluation:** Governing structures frequently require risk evaluations to recognise possible weaknesses and hazards. Metropolises evaluate hazards linked to various intelligent urban elements, aiding them in making knowledgeable choices to prioritise safety expenditures.
- **Responsibility:** Regulatory frameworks establish responsibility for security breaches. They establish obligations, positions, and obligations for various stakeholders, guaranteeing that those engaged in intelligent urban projects are motivated to maintain security benchmarks.
- **Vendor Assessment:** Legislative structures steer the assessment of external suppliers' security methodologies. Metropolises can evaluate merchants based on adherence to established criteria, amplifying acquisition choices and alleviating potential hazards.
- **Flexibility:** Governing structures are crafted to adapt alongside technological progress and emerging dangers. This flexibility guarantees that security measures stay efficient in tackling novel obstacles that emerge over time.

- **Citizen Confidence:** Clearly outlined safety protocols contribute to citizen confidence. When residents are guaranteed that their information is safeguarded and that intelligent urban services are safe, they are more inclined to involve and take part in digital endeavours.
- **Global Uniformity:** Regulatory structures establish a groundwork for worldwide uniformity in intelligent urban centre security measures. Metropolises across the globe can synchronise their security endeavours, rendering it more convenient to exchange encounters, optimal methodologies, and cooperative resolutions.

CONCLUSION

Establishing regulatory frameworks for intelligent metropolis security is vital for ensuring standardised practices that safeguard information, infrastructure, and inhabitants. These structures direct the execution of safety precautions, augment responsibility, and contribute to the formation of protected and durable urban settings. Hui *et al.* (2020) explore the potential of 5G network-based Internet of Things (IoT) for demand response in smart grids. They discuss how IoT sensors and real-time data analytics can enhance energy management, reduce consumption during peak periods, and improve the overall efficiency of smart grids. This source sheds light on the convergence of 5G and IoT in the energy sector.

REFERENCES

[1]   3GPP, "3GPP System Architecture Evolution (SAE); Security architecture," Available from: https://www.3gpp.org/ftp/Specs/archive/33_series/ 33.401

[2]   3GPP, "Architecture enhancements for 5G System (5GS) to support network data analytics services," Available from: https://www.3gpp.org/ftp/Specs/archive/33_series/33.401

[3]   3GPP, "Procedures for the 5G System," Available from: https://portal.3gpp.org/desktopmodules/ Specifications/SpecificationDetails.aspx? specificationId=3145

[4]   3GPP, "Security architecture and procedures for 5G system," Available from: https://portal.3gpp.org/ desktopmodules/Specifications/ SpecificationDetails.aspx?specificationId=3169

[5]   3GPP, "System architecture for the 5G System (5GS)," Available from: https://portal.3gpp.org/ desktopmodules/Specifications/ SpecificationDetails.aspx?specificationId=3144

[6]   I. Ahmad, T. Kumar, M. Liyanage, J. Okwuibe, M. Ylianttila, and A. Gurtov, "Overview of 5G security challenges and solutions", *IEEE Communications Standards Magazine,* vol. 2, no. 1, pp. 36-43, 2018.
[http://dx.doi.org/10.1109/MCOMSTD.2018.1700063]

[7]   F.A. Alaba, M. Othman, I.A.T. Hashem, and F. Alotaibi, "Internet of Things security: A survey", *J. Netw. Comput. Appl.,* vol. 88, pp. 10-28, 2017.
[http://dx.doi.org/10.1016/j.jnca.2017.04.002]

[8]   A. Alhilal, T. Braud, and P. Hui, "Distributed Vehicular Computing at the Dawn of 5G: a Survey," *arXiv,* 2020, Available from: Available from: https://arxiv.org/pdf/2001.07077.pdf

[9]   A. Rghioui, and A. Oumnad, "Internet of things: Surveys for measuring human activities from

everywhere", *International Journal of Electrical and Computer Engineering (IJECE),* vol. 7, no. 5, pp. 2474-2482, 2017.
[http://dx.doi.org/10.11591/ijece.v7i5.pp2474-2482]

[10]   O. Antons, and J.C. Arlinghaus, "Designing decision-making authorities for smart factories", *Procedia CIRP,* vol. 93, pp. 316-322, 2020.
[http://dx.doi.org/10.1016/j.procir.2020.04.047]

[11]   G. Arfaoui, P. Bisson, R. Blom, R. Borgaonkar, H. Englund, E. Felix, F. Klaedtke, P.K. Nakarmi, M. Naslund, P. O'Hanlon, J. Papay, J. Suomalainen, M. Surridge, J-P. Wary, and A. Zahariev, "A Security Architecture for 5G Networks", *IEEE Access,* vol. 6, pp. 22466-22479, 2018.
[http://dx.doi.org/10.1109/ACCESS.2018.2827419]

[12]   J. Cheng, W. Chen, F. Tao, and C.L. Lin, "Industrial IoT in 5G environment towards smart manufacturing", *J. Ind. Inf. Integr.,* vol. 10, pp. 10-19, 2018.
[http://dx.doi.org/10.1016/j.jii.2018.04.001]

[13]   Cisco, Available from: https://www.cisco.com/c/en/us/solutions/collateral/executiveperspectives/annual-internet-report/white-paper-c11-741490.html

[14]   Ericsson, "A guide to 5G network security," Available from: https://www.ericsson.com/en/security/a-guide-to-5g-network-security

[15]   M. Faheem, S.B.H. Shah, R.A. Butt, B. Raza, M. Anwar, M.W. Ashraf, M.A. Ngadi, and V.C. Gungor, "Smart grid communication and information technologies in the perspective of Industry 4.0: Opportunities and challenges", *Comput. Sci. Rev.,* vol. 30, pp. 1-30, 2018.
[http://dx.doi.org/10.1016/j.cosrev.2018.08.001]

[16]   D. Fang, Y. Qian, and R.Q. Hu, "Security for 5G mobile wireless networks", *IEEE Access,* vol. 6, pp. 4850-4874, 2018.
[http://dx.doi.org/10.1109/ACCESS.2017.2779146]

[17]   GSMA, "5G Use Cases for Verticals," Available from: https://www.gsma.com/greater-china/wp content/uploads/2020/03/5G-Use-Cases-for-Verticals-China-2020.pdf

[18]   N.B. Henda, "Overview on the security in 5G Phase 2", *J. ICT Stand,* pp. 1-14, 2020.
[http://dx.doi.org/10.13052/jicts2245-800X.811]

[19]   H. Hui, Y. Ding, Q. Shi, F. Li, Y. Song, and J. Yan, "5G network-based Internet of Things for demand response in smart grid: A survey on application potential", *Appl. Energy,* vol. 257, p. 113972, 2020.
[http://dx.doi.org/10.1016/j.apenergy.2019.113972]

[20]   IETF, "JSON Web Encryption RFC 7516," Available from: https://tools.ietf.org/html/rfc7516

[21]   IETF, "JSON Web Signatures RFC7515," Available from: https://tools.ietf.org/pdf/rfc7515.pdf

[22]   IoT Analytics, "State of the IoT 2018: Number of IoT devices now at 7B," Available from: https://iot-analytics.com/state-of-the-iot-update-q1-q2-2018-number-of-iotdevices-now-7b/

[23]   Kaspersky, "Kaspersky detects more than 100 million attacks on smart devices in H1 2019," Available from: https://www.kaspersky.com/about/pressreleases/2019_iot-under-fire-kaspersky-detects-more-than-100-million-attackson-smart-devices-in-h1-2019

[24]   J.S. Kho, and J. Jeong, "HACCP-based cooperative model for smart factory in South Korea", *Procedia Comput. Sci.,* vol. 175, pp. 778-783, 2020.
[http://dx.doi.org/10.1016/j.procs.2020.07.116] [PMID: 32834881]

[25]   J. Kim, G. Jo, and J. Jeong, "A novel CPPS architecture integrated with centralized OPC UA server for 5G-based smart manufacturing", *Procedia Comput. Sci.,* vol. 155, pp. 113-120, 2019.
[http://dx.doi.org/10.1016/j.procs.2019.08.019]

[26]   K. Kimani, V. Oduol, and K. Langat, "Cyber security challenges for IoT-based smart grid networks", *Int. J. Crit. Infrastruct. Prot.,* vol. 25, pp. 36-49, 2019.
[http://dx.doi.org/10.1016/j.ijcip.2019.01.001]

[27] Y. Levy, and T. Ellis, "A systems approach to conduct an effective literature review in support of information systems research", *Inf. Sci.,* vol. 9, pp. 181-212, 2006.
[http://dx.doi.org/10.28945/479]

[28] S. Li, L.D. Xu, and S. Zhao, "5G internet of things: A survey", *J. Ind. Inf. Integr.,* vol. 10, pp. 1-9, 2018.
[http://dx.doi.org/10.1016/j.jii.2018.01.005]

[29] Nokia, "Nokia Bell Labs Industrial Automation Networks," Available from: https://www.nokia.com/networks/training/5g/bell-labs/courses/

[30] M. Pauliac, "USIM in 5G Era", *J. ICT Stand,* pp. 29-40, 2020.
[http://dx.doi.org/10.13052/jicts2245-800X.813]

[31] K.M. Sadique, R. Rahmani, and P. Johannesson, "Towards security on internet of things: Applications and challenges in technology", *Procedia Comput. Sci.,* vol. 100, pp. 872-879, 2016.
[http://dx.doi.org/10.1016/j.procs.2016.09.144]

[32] W. Zhang, H. Ren, W. Chen, H. Zheng, and Y. Yang, "Security for 5G networks: Threats and countermeasures", *Comput. Sci. Rev.,* vol. 31, pp. 1-16, 2019.
[http://dx.doi.org/10.1016/j.cosrev.2019.03.001]

[33] A.K. Sharma, P.G.S.N. Reddy, S.M.R. Guda, and S.K. Ghosh, "Smart grid technology and its applications for future power systems", *Energy Rep.,* vol. 6, pp. 359-366, 2020.
[http://dx.doi.org/10.1016/j.egyr.2020.05.025]

[34] A.S. Siddiqui, M.B. Anwar, A.A. Shaikh, and S. Zeadally, "Security challenges in vehicular ad hoc networks", *Ad Hoc Netw.,* vol. 35, pp. 87-102, 2015.
[http://dx.doi.org/10.1016/j.adhoc.2015.03.004]

[35] J.B. Silva, J.F.S. Lima, and P.S.R. Diniz, "A survey on the design of secure communication protocols in V2X networks", *Ad Hoc Netw.,* vol. 88, pp. 88-103, 2019.
[http://dx.doi.org/10.1016/j.adhoc.2019.06.003]

[36] A.S.T. Toure, L.T.T. Pham, and H.K.A. Nguyen, "Data security in Internet of Vehicles for the future smart transportation systems", *Transp. Res., Part C Emerg. Technol.,* vol. 106, pp. 183-198, 2019.
[http://dx.doi.org/10.1016/j.trc.2019.07.012]

[37] M. Xu, P. Liu, L. Zhang, Z. Li, and X. Liu, "A comprehensive survey of V2X security in 5G networks", *Comput. Commun.,* vol. 147, pp. 46-62, 2019.
[http://dx.doi.org/10.1016/j.comcom.2019.10.009]

[38] Y. Zhang, Z. Chen, and T. Wang, "5G V2X communications for autonomous driving: Architecture, technologies, and challenges", *IEEE Trans. Industr. Inform.,* vol. 15, no. 4, pp. 2396-2403, 2019.
[http://dx.doi.org/10.1109/tii.2018.2854563]

[39] A.F. Zolkipli, M.Z.A. Raja, H. Shamsuddin, and M.I.F.A. Ibrahim, "Security and privacy issues for connected vehicles", *Comput. Netw.,* vol. 162, pp. 25-38, 2019.
[http://dx.doi.org/10.1016/j.comnet.2019.06.007]

[40] X. Zhang, and C.C. Ko, C. A. G. P. K. R. S. M. Z., and H. Yang, "Design and optimization of a V2X-based cooperative driving system for autonomous vehicles", *IEEE Trans. Intell. Transp. Syst.,* vol. 21, no. 3, pp. 1142-1151, 2020.
[http://dx.doi.org/10.1109/its.2019.2950679]

[41] D. Nataraju, D. Pradhan, and S.S. Jambli, "Opportunities, challenges, and benefits of 5G-IoT toward sustainable development of green smart cities (SD-GSC)", *3rd International Conference on Intelligent Technologies (CONIT),* pp. 1-8, 2023.

[42] D. Pradhan, R. Suma, R. Carva, and C.B. Carva, "Implications of resource optimization in D2D communication for efficient usage of RF energy: Practice and challenges", *3rd International Conference on Intelligent Technologies (CONIT),* pp. 1-8, 2023.

[http://dx.doi.org/10.1109/CONIT59222.2023.10205729]

[43]   H.K. Sinha, D. Pradhan, S.A. Saurabh, and A. Kumar, Software principles of 5G coverage: Simulator analysis of various parameters.*The Software Principles of Design for Data Modeling.* IGI Global, 2023, pp. 276-285.
[http://dx.doi.org/10.4018/978-1-6684-9809-5.ch020]

[44]   H.K. Sinha, A. Kumar, and D. Pradhan, A study of various peak to average power ratio (PAPR) reduction techniques for 5G communication system (5G-CS)*Optimization Techniques in Engineering: Advances and Applications,* pp. 437-454, 2023.
[http://dx.doi.org/10.1002/9781119906391.ch27]

[45]   D. Pradhan, P.K. Sahu, H.M. Tun, and N.K. Wah, Integration of AI/ML in 5G technology toward intelligent connectivity, security, and challenges.*Machine Learning Algorithms and Applications in Engineering.* CRC Press, 2023, pp. 239-254.
[http://dx.doi.org/10.1201/9781003104858-14]

[46]   P.K. Priyanka, G. Mallavaram, A. Raj, D. Pradhan, and R.S. Rajeswari, Cognitiveness of 5G technology toward sustainable development of smart cities*Decision Support Systems for Smart City Applications,* pp. 189-203, 2022.
[http://dx.doi.org/10.1002/9781119896951.ch11]

CHAPTER 4

5G and Smart Cities: Smarter Solutions for a Hyperconnected Future

Rakesh Kumar Dixit[1,*] and **Pushpendra Pal Singh**[1]

[1] *G.L. Bajaj Institute of Management, Greater Noida, India*

Abstract: The integration of 5G technology into the fabric of smart cities heralds a new era of urban development, promising unprecedented levels of connectivity, efficiency, and innovation. This paper explores the transformative potential of 5G networks in shaping the future of smart cities, where hyperconnectivity serves as the cornerstone for smarter solutions to address pressing urban challenges. Beginning with an overview of the fundamental principles underlying smart cities, this paper highlights the imperative of leveraging advanced technologies to create more sustainable, resilient, and livable urban environments. It then examines the unique capabilities of 5G networks, including ultra-fast data transmission, ultra-low latency, and massive device connectivity, and explores how these features enable a diverse array of smart city applications across various sectors. Furthermore, the paper delves into the specific ways in which 5G technology enhances existing smart city infrastructures and enables the development of novel solutions to urban challenges. From intelligent transportation systems and autonomous vehicles to remote healthcare services and augmented reality experiences, the hyperconnectivity facilitated by 5G networks empowers cities to deploy innovative solutions that improve quality of life for residents and enhance urban efficiency.Moreover, the paper discusses the challenges and opportunities associated with the deployment of 5G networks in urban environments, including infrastructure requirements, regulatory considerations, and privacy concerns. It emphasizes the need for collaboration between governments, industries, and communities to address these challenges and ensure the responsible and equitable deployment of 5G technology in smart cities.

Keywords: Edge computing, Hyperconnected networks, IoT (internet of things), Network slicing, Smart cities, Urban innovation, 5G technology.

INTRODUCTION

Throughout the span of approximately every decade, a novel surge of wireless mobile telecommunications technology arises, characterised by the application of

---

* **Corresponding author Rakesh Kumar Dixit:** G.L. Bajaj Institute of Management, Greater Noida, India;
E-mail: rakeshdixit578@gmail.com

Devasis Pradhan, Mangesh M. Ghonge, Nitin S. Goje, Alessandro Bruno and Rajeswari (Eds.)

inventive frequency bands, elevated data speeds, and the inception of brand-new services. This advancement propels us nearer to accomplishing the smooth connectivity of almost every aspect of our tangible world.

The inaugural phase, known as 1G, made its debut in the early 1980s. It was distinguished by its capacity to transmit speech using analogue technology. While exemplifying noteworthy advancement for its era, 1G had constraints. Significantly, it lacked data services to transform voice into digital signals, displayed below-average voice quality, and had not yet provided global roaming services.

In this comprehensive review, the (Digital 2020) report provides detailed insights into global digital trends. It covers a wide range of digital indicators, including internet penetration, social media usage, and e-commerce trends, offering a holistic view of the digital landscape that can inform discussions on technology adoption.

The ensuing era, 2G, emerged in the tardy 1990s, signifying the onset of digital technology. 2G brought improvements in vocal clarity and data transfer capability. In this period, the Worldwide System for Mobile Communications (WSMC) acted as the digital norm, integrating characteristics such as Brief Communication Assistance (BCA), Multimedia Communication Assistance (MCA) enabled by devices with vivid screens, and Wireless Application Protocol (WAP) enabling internet connectivity services on portable gadgets. Notwithstanding the energy-intensive characteristic of these multimedia applications, a noteworthy benefit of 2G mobile devices was their prolonged battery life, attributable to the minimal power usage of radio signals [1].

The third stage, 3G, surfaced in the latter portion of the 2000s. It introduced authentic wireless data connectivity, granting users extensive internet access. Significantly, 3G technology brought about heightened data transmission velocities, enabling the evolution of advanced multimedia applications. Moreover, the incorporation of new frequency ranges and positioning data allowed for functionalities previously unavailable to portable devices. This included endeavours like web surfing, electronic mail entry, TV streaming, visual assembly, and even the application of global positioning system (GPS) technology. This expansive assortment of applications made the 3G era remarkable for the consumer market. Nevertheless, it additionally resulted in a surge in expenditures for 3G gadgets and amplified power usage. For instance, 3G devices require more power compared to the majority of 2G models due to their expanded capabilities [2].

The fourth stage of wireless mobile telecommunications technology, recognised as 4G, surfaced in 2010 and persists to be employed presently. This era is constructed upon Internet Protocol (IP) and strives to offer exceptional, safeguarded, economical services, multimedia, and internet connectivity *via* IP, presenting notably elevated data speeds in comparison to its forerunners. Precisely, 4G introduces pervasive high-speed wireless broadband, unleashing the potential of portable video streaming and advanced services like interactive amusement, high-definition streaming, and three-dimensional television [3].

As utilities increasingly turn to drones for inspections and maintenance, a review like [4] can provide insights into the adoption and impact of drone technology in the utility sector.

Currently, 4G provides consumer data speeds in the magnitude of megabytes, latency within the millisecond spectrum, and sustains a device density of approximately 2000 connected devices per square kilometre worldwide. This has enabled the execution of the Internet of Things (IoT). Nevertheless, owing to an escalating requirement and the advent of innovative cellular communication breakthroughs, the 4G epoch is anticipated to pave the way for the subsequent iteration, 5G, at the onset of the forthcoming decade [5].

The forthcoming 5G epoch is ready to introduce network and service capabilities formerly unreachable. It assures improved durability, intensified data speeds, decreased delay, backing for extensive simultaneous links, and worldwide network reach even in demanding situations like high mobility (*e.g.*, in trains) and densely packed or thinly populated areas (*e.g.*, stadiums, marketplaces). Furthermore, 5G will have a crucial function in facilitating an authentic Internet of Things (IoT), offering a groundwork to link a vast assortment of detectors and effectors with an emphasis on energy effectiveness and transmission constraints.

Motivated by an unparalleled upsurge in interconnected gadgets, mobile data traffic, and the constraints of 4G technology in tackling this substantial data requirement, both industry and academia are concentrated on delineating the specifications for 5G services, signifying the commencement of the 5G epoch. A contraption equipped with 5G will possess the ability to sustain network connectivity consistently and universally, unleashing the potential to interconnect all contraptions within the network. To accomplish this objective, the fundamental plan of the 5G system is expected to endorse up to a million simultaneous connections per square kilometre, enabling the actualization of a plethora of inventive concepts within the domain of Internet of Things (IoT) amenities [6].

The Internet of Things embodies a modern digital communication paradigm, where everyday objects interact with each other and users through the World

Wide Web. Therefore, IoT aspires to expand the notion of the Internet, enhancing its immersive nature by enabling smooth interactions with a wide array of gadgets like domestic apparatus, monitoring devices, manufacturing equipment, traffic indicators, and vehicles, amidst other things [7]. In this scenery, an abundance of information is produced and gathered from the vast web of interconnected devices. The incorporation of Cloud Computing and Big Data technologies plays a noteworthy function in handling various data categories in accordance with requirements, producing progressively valuable services. These technologies are indispensable for realizing the IoT paradigm in urban contexts, often referred to as Intelligent Cities [8]. This approach accommodates the necessity of national governments to embrace Information and Communications Technologies (ICT) solutions for efficient governance of public affairs shown in Fig. (**1**).

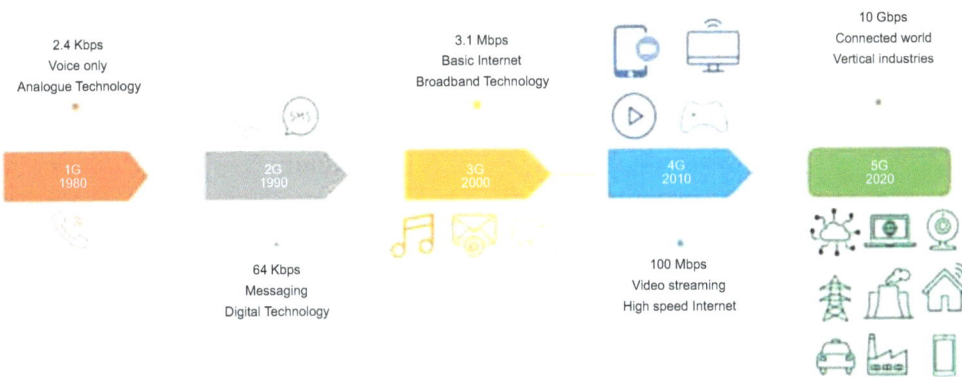

Fig. (1). The evolution of mobile communications [1].

SMART CITIES AND GREEN TECHNOLOGY

In the milieu of swift urbanisation and escalating environmental predicaments, the notion of intelligent cities has materialised as a guiding light of sustainable urban advancement. These groundbreaking metropolitan habitats utilise cutting-edge technologies to improve the quality of existence while reducing the ecological footprint. At the core of this metamorphosis lies the fifth-gen (5G) network technology, ready to overhaul how cities function and propel the assimilation of eco-friendly technology. This amalgamation of intelligent metropolises and 5G networks harbours vast potential in optimising resource utilisation, enhancing energy effectiveness, augmenting ecological surveillance, and transforming urban transportation. As metropolitan areas progressively aspire for ecological sustainability, the mutually beneficial connection between intelligent cities and 5G networks stands as evidence of the capacity for technology to cultivate more environmentally friendly, robust urban environments [9].

Smart Cities: Concepts and Characteristics

Intelligent cities epitomise a revolutionary strategy to metropolitan growth, leveraging technological advancement to address the intricate dilemmas presented by urbanisation and ecological durability. At their essence, intelligent cities are constructed upon the smooth integration of technology, information, and amenities, striving to amplify the welfare of inhabitants, advocate ecological guardianship, and optimise resource effectiveness [10].

Definition and Core Components of Smart Cities: Smart cities can be defined as urban ecosystems that utilize advanced technologies to interconnect various aspects of urban life, including infrastructure, public services, and communication networks. These cities leverage data-driven insights to enhance operational efficiency, promote citizen engagement, and address urban challenges proactively.

The core components of smart cities encompass:

Intelligent Infrastructure: Clever cities deploy state-of-the-art technologies to construct astute infrastructure that reacts dynamically to urban needs. This encompasses intelligent networks, water administration systems, and energy-conscious structures engineered to maximise resource utilisation and reduce inefficiency [11].

Data Fusion: At the core of intelligent cities lies the gathering, examination, and exploitation of information from diverse origins, including detectors, portable gadgets, and online platforms. By consolidating and scrutinising this data, cities acquire actionable perspectives that empower knowledgeable decision-making across varied sectors.

Connectivity: Intelligent cities prioritise rapid connectivity through broadband networks and wireless technologies like 5G. This interconnectivity establishes the foundation of instantaneous communication, facilitating smooth exchanges between individuals, gadgets, and metropolitan amenities.

Integration of Technology, Data, and Services: The characteristic of intelligent cities is the incorporation of technology, information, and amenities into a consolidated environment. This amalgamation enables the formation of receptive metropolitan systems that adjust to fluctuating circumstances and amplify the calibre of existence for inhabitants. Cutting-edge technologies like the Internet of Things (IoT) facilitate the smooth interaction of gadgets and detectors, producing a plethora of information that is employed to enhance services and urban operations [12].

Focus on Improving Quality of Life, Sustainability, and Resource Efficiency Intelligent cities prioritise enhancing the standard of living for their inhabitants. This entails offering effective and reachable public services, nurturing a feeling of camaraderie, and augmenting security and welfare. Furthermore, the incorporation of eco-friendly measures is crucial, with an emphasis on diminishing energy usage, mitigating contamination, and preserving ecological assets [13].

Attempts to improve resource effectiveness are likewise fundamental to the intelligent urban philosophy. By means of data-fueled insights, cities can optimise transportation networks, streamline waste management procedures, and implement energy-conscious practises that together diminish the ecological impact of urban living.In summary, intelligent cities embody a comprehensive strategy for urban advancement that utilises technology, information, and amenities to establish a greener and more effective urban setting. By incorporating clever infrastructure, data-powered decision-making, and swift connectivity, intelligent cities aspire to enhance the standard of living for inhabitants while tackling ecological difficulties and promoting conscientious resource administration [15].

Green Technology in Smart Cities

In the quest for sustainable urban development, the incorporation of eco-friendly technology has arisen as a vital aspect within the structure of intelligent cities. Eco-friendly technology, frequently known as environmentally friendly technology or sustainable technology, encompasses inventive solutions that alleviate ecological influence while advocating for resource effectiveness and preservation [16]. Within intelligent cities, the integration of eco-friendly technology plays a crucial role in achieving the goal of ecologically sustainable urban living.

Definition of Green Technology and Its Objectives: Sustainable technology can be described as the utilisation of scientific and engineering principles to generate goods, procedures, and frameworks that reduce adverse ecological impacts. Its main goals encompass:

Mitigating Ecological Footprint: Sustainable technology aims to diminish or eradicate the detrimental consequences of human actions on the ecosystem. This comprises different facets, encompassing energy usage, refuse production, and contamination.

Preserving Resources: Through maximising resource utilisation and reducing waste generation, eco-friendly technology strives to prolong the longevity of limited resources and diminish the environmental impact of human actions.

Promoting Sustainability: Eco-friendly technology endeavours to establish solutions that fulfil current requirements without jeopardising the capacity of future generations to fulfil their own requirements. It corresponds with the wider objectives of sustainable development and enduring ecological conservation [17]. The concept of private 5G networks for industrial IoT applications. The study explores the benefits of private networks in enhancing connectivity, reliability, and security within industrial settings [18].

Role of Green Technology in Achieving Environmental Sustainability

Within the framework of intelligent cities, eco-friendly technology acts as a pivotal point for attaining ecological sustainability. By incorporating ecologically conscious practises and advancements into urban systems, sustainable technology contributes to:

Energy Conservation: Eco-friendly technology solutions like energy-conserving illumination, intelligent networks, and sustainable energy sources diminish energy usage and emissions of greenhouse gases, thereby lessening the ecological impact of urban areas.

Waste Administration: Progressive waste administration technologies, such as reusing initiatives and waste-to-power conversion, advocate conscientious waste disposal and diminish landfill waste, lessening ecological damage.

Water Preservation: Eco-friendly technology implementations in water administration, encompassing precipitation collection and intelligent watering systems, enhance water utilisation and alleviate pressure on nearby water reserves.

Air Quality Enhancement: Execution of electric communal transport systems, alongside enhanced surveillance of air quality through sensors, contributes to purer air and diminished discharges in metropolitan regions.

Biodiversity Conservation: Verdant areas, upright gardens, and metropolitan agriculture initiatives nurture biodiversity within urban settings, augmenting ecological resilience and enhancing overall urban aesthetics [19].

Examples of Green Technology Applications in Urban Settings

Sustainable technology discovers functional implementations throughout diverse metropolitan industries:

Sustainable Energy: Photovoltaic modules, windmills, and hydroelectric installations produce eco-friendly energy to fuel metropolitan infrastructure.

Intelligent Structures: Energy-savvy architectural plans, intelligent temperature regulators, and automated illumination systems optimise energy consumption within edifices [20].

Electric Mobility: Electric automobiles, electric recharging stations, and intelligent transportation systems diminish atmospheric contamination and reliance on non-renewable resources.

Waste-to-Power: Techniques that transform biodegradable waste into energy contribute to both waste minimization and energy production.

Intelligent Aquatic Management: Intelligent gauges, seepage identification frameworks, and water filtration advancements enhance water preservation and excellence [21].

Fundamentally, the incorporation of eco-friendly technology within the structure of intelligent cities enhances the quest for sustainable urban existence. By synchronising technological advancements with ecological accountability, eco-friendly technology converts cities into centres of ingenuity, effectiveness, and environmental guardianship.

5G TECHNOLOGY: ENABLING THE SMART CITY ECOSYSTEM

Pivotal to the metamorphosis of intelligent metropolises into vibrant and linked urban centres is the groundbreaking fifth-generation (5G) technology. With its unparalleled capabilities, 5G technology acts as the catalyst that propels the advancement of intelligent urban environments. By offering super-swift connectivity, minimal delay, and the capability to accommodate a wide array of gadgets, 5G technology unleashes the potential for a fresh epoch of urban development that is information-based, adaptable, and eco-friendly [22]. This segment delves into how 5G technology empowers the intricate network of intelligent urban components, ranging from IoT devices and live data analysis to smart transport systems and citizen involvement platforms. As metropolises embrace 5G technology, they position themselves at the vanguard of a revolutionary voyage towards more effective, linked, and eco-friendly urban existence shown in Fig. (**2**).

Penttinen [23], offers an in-depth exploration of 5G technology, focusing on its positioning within the context of smart cities. The study delves into the various applications and implications of 5G networks in transforming urban environments, including enhanced connectivity, IoT integration, and improved mobile communications.

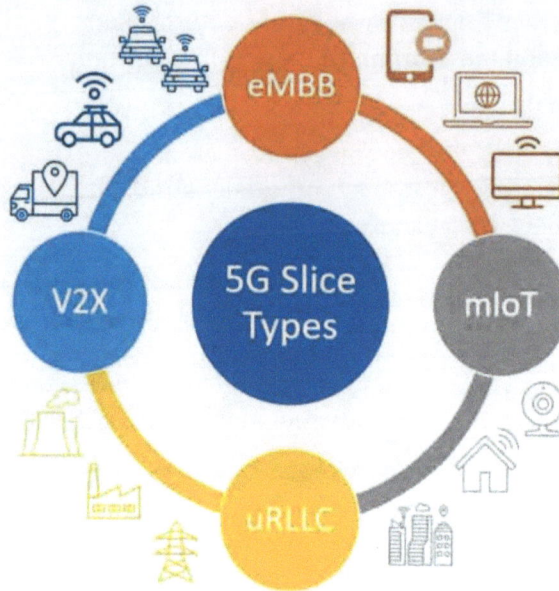

Fig. (2). Slice types supported by 5G [2].

Understanding 5G Networks

The inception of fifth-generation (5G) technology signifies a crucial moment in the progression of wireless communication systems. 5G technology embodies a paradigm change, introducing a variety of capabilities that greatly exceed its forerunners and empowering the achievement of the intelligent metropolis ecosystem.

Explanation of 5G Technology and Its Key Features: At its essence, 5G technology constructs upon the bedrock of preceding wireless generations while introducing revolutionary improvements. It functions on elevated frequency bands, utilising reduced wavelengths to accomplish superior data velocities and broader data capability. This escalated capability is a foundation of 5G's aptitude to sustain the varied and information-heavy functionalities emblematic of intelligent metropolises.

High Data Speeds, Low Latency, and Massive Device Connectivity

Noteworthy characteristics of 5G technology encompass its extraordinary data velocities, extremely minimal delay, and the ability to link an unparalleled multitude of devices concurrently. Data velocities in the span of gigabits per second enable nearly instantaneous downloads and uploads, improving real-time communication and data transmission. Minimal delay, frequently assessed in milliseconds, guarantees that devices can communicate without noticeable pauses, a pivotal element for applications like self-driving cars and distant medical procedures.

Arguably, the utmost noteworthy aspect of intelligent cities is the capacity of 5G to accommodate an immense multitude of gadgets within a specified vicinity. This capability, propelled by cutting-edge antenna technologies, clears the path for the spread of Internet of Things (IoT) devices that collect and transmit data from diverse urban systems. The uninterrupted interconnectedness empowered by 5G facilitates the interchange of information between sensors, vehicles, infrastructure, and citizens, constructing the foundation of the intelligent city ecosystem.

Network Slicing and Edge Computing Capabilities

A novelty introduced by 5G, network partitioning involves the separation of a tangible network into numerous virtual networks, each customised to particular applications or user clusters. This adaptability enables intelligent cities to assign resources accurately where required, optimising network efficiency for different services and ensuring effective resource utilisation.

Furthermore, the edge computing capabilities of 5G are pivotal for intelligent urban areas. Edge computing entails the processing of data in proximity to its origin instead of transmitting it to a remote data hub. This diminishes delay and amplifies instantaneous processing, enabling applications like self-driving cars to make instantaneous judgements based on nearby information.

In summary, comprehending 5G networks is crucial for grasping the transformative capacity they bring to intelligent cities. The amalgamation of elevated data velocities, diminished latency, immense device connectivity, network segmentation, and periphery computation capabilities empower cities to fabricate a vibrant and interlinked ecosystem. As intelligent cities utilise these technological progressions, they unlock the gateway to inventive amenities, streamlined asset administration, and a future that embraces the complete capability of the digital era.

5G and IoT Integration

The amalgamation of fifth-generation (5G) technology and the Internet of Things (IoT) possesses revolutionary potential for shaping the panorama of intelligent metropolises. As intelligent metropolises develop into elaborate ecosystems of interconnected gadgets and amenities, the incorporation of 5G and IoT emerges as a pivotal point for attaining effective resource administration and enriching urban existence.

Internet of Things (IoT) and Its Role in Smart Cities: The Internet of Things (IoT) alludes to the extensive web of interconnected gadgets and detectors that gather and transmit information, cultivating a mutually beneficial association between the virtual and tangible realms. In intelligent cities, the Internet of Things acts as a vital structure for collecting up-to-the-minute data about diverse urban elements, such as power usage, atmospheric conditions, movement of vehicles, and garbage control. This information-based approach enables urban managers to make knowledgeable choices and implement focused interventions that optimise resource distribution and improve the overall quality of living.

How 5G Enhances IoT Connectivity and Communication

The inception of 5G technology transforms the capabilities of the IoT within intelligent cities. Conventional wireless networks frequently face challenges in accommodating the immense amount of data produced by the Internet of Things (IoT) devices. Nevertheless, 5G's elevated data velocities, reduced latency, and extensive device connectivity tackle these constraints by furnishing the essential framework for uninterrupted communication amidst devices, sensors, and central systems.

5G's enhanced bandwidth and capability enable the concurrent connection of a multitude of devices, guaranteeing that the IoT can function at its maximum capacity. The diminished latency provided by 5G allows instantaneous communication, crucial for applications that require momentary responsiveness, like self-driving cars and distant medical operations. This collaboration between 5G and IoT enables intelligent cities to achieve their data-focused aspirations, transforming immediate observations into implementable resolutions [24].

Enabling Diverse IoT Applications for Efficient Resource Management

The incorporation of 5G and IoT unveils a wide range of applications that contribute to effective resource administration in intelligent cities.

Energy Maximisation: 5G-facilitated intelligent networks oversee energy consumption in live moments, enabling flexible allocation and diminishing inefficiency.

Waste Management: IoT detectors in garbage bins transmit capacity information, allowing optimised collection routes and reducing unnecessary pickups.

Air Quality Surveillance: 5G-linked detectors offer uninterrupted air quality information, enabling cities to recognise contamination origins and execute adaptive actions.

Traffic Control: 5G-fueled traffic detectors facilitate live tracking of traffic movement, enabling flexible traffic administration systems to alleviate gridlock.

Public Security: IoT-enabled surveillance cameras and sensors amplify public security measures, providing live monitoring of urban areas.

In substance, the amalgamation of 5G and IoT is a foundation of the intelligent metropolis uprising. By augmenting connectivity, correspondence, and information interchange, this collaboration empowers intelligent cities to harness up-to-the-minute perceptions for effective resource administration and the establishment of more habitable, eco-friendly urban settings.

5G FOR SMART CITIES

Intelligent cities are targeted at enhancing the utilisation of communal resources, amplifying the calibre of the amenities with emphasis on convenience, upkeep, and durability, while diminishing the operational expenses of the communal facilities, within an Internet of Things (IoT) framework. Individual and Household Applications are the initial classification and encompass domestic devices linked and pervasive e-healthcare services that aid physicians in remotely monitoring patients. Utility Application is the subsequent classification and encompasses intelligent water network monitoring, atmospheric condition, video-centric surveillance, communal security and exigency provision. The third classification is Industrial Implementations, which typically comprises a network of industrial apparatus within a manufacturing setting. The final classification is focused on Smart Transportation Systems (STS) or broadly, Mobility Solutions. The subsequent classification encompasses burgeoning notions such as self-driving cars, automobile networks, traffic administration, and gridlock regulation, amidst other alternatives [25].

Numerous investigation endeavours have been undertaken to amalgamate 5G technologies and IoT services in intelligent urban settings, a few of them energised by industries and others by academia. Herein, we will present some of the most pertinent methodologies within each application category emphasising mobility applications [26].

Personal and Home Applications

Currently, a diminutive proportion of individuals possess a wellness gadget alternatively referred to as a label device, but the possibilities are extensive and with 5G, intelligent label devices are anticipated to become more widespread. In contrast to contemporary gadgets, upcoming 5G devices will be entirely linked as there will be no requirement to be tethered to a mobile device for internet connectivity. Companies like Samsung are advancing healthcare and wellness equipment that not only document physical activity accomplishments and provide suggestions regarding workout regimens, but also transmit imperative healthcare data to a specialist instantaneously to avert or supervise medical crises [27].

Utilities Applications

Utilising 5G for urban IoT could potentially offer surveillance of the entire energy usage in the city, thereby empowering authorities to obtain comprehensive and precious data regarding the energy needed by various public amenities (*e.g.,* public illumination, traffic signals, security cameras, heating/cooling of public structures, among other things). This will enable recognising the primary energy consumption sources and subsequently strategizing to enhance urban energy administration [28].

Furthermore, apart from the financial advantage of optimising energy resources, 5G is anticipated to aid public safety by preserving lives through calamity and urgency retort or enhancing crime identification and surveillance. Dubious luggage in airports, defacement and delinquent recognition can be countered by employing monitoring cameras and computational perception methods. When the threat is identified, utilising 5G rapid connection, this will be conveyed to public safety personnel who could aid in synchronising response measures. Furthermore, a security system is showcased that, upon identifying the appearance of a familiar wrongdoer, even if the offence has not yet transpired, the system seizes real-time images and precise locations to be forwarded to the neighbouring law enforcement facilities.

Industrial Applications

The rapid surge of contemporary technologies (comprising immense data, cloud computing, synthetic intelligence, and 5G) has captivated significant attention from the industry to incorporate ICT in the manufacturing setting. The fusion of industrial machinery with ICT unveils possibilities to expedite productivity, diminish excess, enhance efficiency, and ameliorate the working experience in the manufacturing milieu. Farming is a distinct domain where IoT has immense potential. Utilising detectors with wireless connectivity for agricultural fields can aid in optimising cultivation and reducing the utilisation of moisture and nutrients. Cattle, reservoirs and other agricultural machinery can be observed from a distance, enhancing farming efficiency by diminishing manufacturing expenses.

Mobility Applications

Progressively, metropolitan automobiles are evolving into a mobile sensor platform that furnishes ecological data to drivers and in the near future, such data could be uploaded to the cloud. The detectors' information will be accessible to a network of self-driving cars that share their data with one another to enhance a clearly defined purpose. Hence, automobiles would transform into an additional gadget linked to the Internet.

Ideally, when human command is eliminated, self-driving vehicles should collaborate to enable managing traffic more proficiently, with reduced wait times, decreased emissions, and enhanced driver and passenger convenience. For example, for catastrophe administration, the automotive network should have the capacity to synchronise the evacuation of hazardous regions in a prompt and organised fashion. This necessitates being capable of communicating with one another additionally possessing access to resources such as emergency vehicles, law enforcement automobiles, or knowledge regarding getaway pathways [29].

For a comprehensive understanding of machine learning techniques in cellular networks, Klaine *et al.* [30], present an in-depth survey. The study covers various machine learning approaches applied to self-organizing cellular networks, discussing their potential for optimizing network performance and management.

However, owing to the intricacy of concurrent management of hundreds of thousands of vehicles, present 4G technologies are incapable of sustaining such a substantial device density. Some additional crucial aspects such as delay and level of service are essential to accomplish it. For example, it would require approximately 1.5 metres for a vehicle with 4G to engage its brakes. An automobile with 5G would solely necessitate 2.5 centimetres to accomplish this, aiding in evading mishaps. Similarly, when a vehicle ventures into a region with limited signal or is highly congested, a 4G connection falters. Nevertheless, a 5G connection hypothetically will perpetually possess coverage, enabling maintaining a steadfast connection everywhere and at any given moment.

Joss *et al.* [31], conduct a detailed examination of the concept of smart cities as a global discourse. The study investigates the storylines and critical junctures across 27 cities, shedding light on the diverse narratives and trajectories of smart city development.

Hence, in the pursuit of IoT and intelligent metropolises, automobiles assume a crucial function that results in the Internet of Automobiles (IoA), which is not solely focused on the interchange among vehicles, but also on individuals, urban areas, or even nations. In this setting, the remaining part encompasses the model of intelligent urban areas concerning automotive communications, displaying the constraints of present-day technologies and essential prerequisites of 5G for Intelligent Transportation Systems (ITS), along with the influence on the ecosystem and the community of notions like ingenious guidance and self-driving.

TRANSFORMING URBAN MOBILITY AND TRANSPORTATION

The amalgamation of intelligent metropolis endeavours and fifth-gen (5G) technology has sparked a revolutionary alteration in the domain of urban mobility and transportation. As metropolitan areas wrestle with traffic gridlock, atmospheric contamination, and ineffective transportation systems, the incorporation of intelligent technologies and fifth-generation networks arises as a stimulant for transforming how individuals and commodities traverse within urban areas. This incorporation empowers the implementation of intelligent transport systems, interconnected vehicles, and information-based traffic management solutions, reshaping urban mobility into a more sustainable, proficient, and receptive ecosystem. As metropolises utilise the potential of 5G technology to facilitate uninterrupted correspondence among automobiles, infrastructure, and inhabitants, the concept of a more secure, eco-friendly, and interconnected urban transportation system emerges as an achievable actuality.

Intelligent Transportation Systems (ITS)

As urbanisation hastens, the predicaments of traffic gridlock, ecological contamination, and ineffective transport systems have become urgent problems. In this particular setting, the incorporation of intelligent transport systems (ITS) and the potential of fifth-generation (5G) technology presents a hopeful resolution to revolutionise urban mobility.

Current Challenges in Urban Transportation: Conventional metropolitan transport systems struggle with gridlock, leading to fruitless time, more fuel usage, and intensified atmospheric contamination. Moreover, insufficient infrastructure and subpar traffic control worsen these difficulties, impeding the standard of living for city residents. Resolving these concerns requires a paradigm alteration that utilises technology to establish a more effective, enduring, and interconnected transportation ecosystem.

Role of 5G in Creating Efficient and Connected Transportation Systems

The implementation of 5G technology in urban transport signals a novel epoch of mobility solutions that prioritise effectiveness and connectivity. With its elevated data velocities and minimal latency, 5G presents a dependable and immediate communication framework that enables real-time interactions amidst vehicles, infrastructure, and central control systems. This connection establishes the foundation of ITS, allowing the execution of dynamic traffic control tactics, anticipatory upkeep, and information-based decision-making (3GPP, 2020).

Real-Time Traffic Management, Smart Parking, and Vehicle-to-Everything (V2X) Communication 5G-capable ITS unveils a variety of revolutionary applications:

Real-Time Traffic Control: With real-time information exchange enabled by 5G, traffic control systems can oversee and react to evolving traffic circumstances on the go. This permits flexible traffic signal timing, redirecting of vehicles to evade gridlock, and dynamic administration of road networks.

Intelligent Parking: 5G technology facilitates the implementation of intelligent parking systems that direct drivers to accessible parking spots. This lessens gridlock caused by motorists seeking parking, resulting in enhanced space utilisation and diminished emissions.

Vehicle-to-Everything (V2X) Communication: 5G's minimal delay and elevated dependability facilitate uninterrupted communication between automobiles, walkers, and infrastructure (V2X). This correspondence improves road safety by notifying drivers of potential crashes, facilitating synchronised traffic movement, and optimising pedestrian crossing durations (ITU, 2015).

In summary, the incorporation of 5G technology and intelligent transport systems has the capability to transform urban mobility. By tackling present transportation obstacles *via* live traffic control, intelligent parking solutions, and V2X communication, cities can establish a transportation ecosystem that is more effective, linked, and ecologically sustainable. This amalgamation contributes to a more radiant future of metropolitan mobility that prioritises security, effectiveness, and an elevated standard of living for all inhabitants [32].

Electric and Autonomous Vehicles

The convergence of electric mobility and self-driving cars with fifth-generation (5G) technology has introduced a revolutionary epoch in urban transport, pledging to reshape cities' mobility panorama towards sustainability and effectiveness.

Promoting Electric Mobility for Reduced Emissions: Electric Vehicles (EVs) have surfaced as a foundation of environmentally conscious transportation solutions. By substituting internal combustion engines with electric motors, EVs greatly diminish emissions, enhance air quality, and alleviate the consequences of climate change. The incorporation of 5G technology in electric mobility additionally improves this shift by facilitating effective charging infrastructure, enabling live vehicle monitoring, and optimising energy usage.

5G Support for Autonomous Vehicle Communication and Navigation: Self-driving automobiles, fueled by synthetic intelligence and detectors, possess the capability to transform metropolitan transportation. The incorporation of 5G technology offers the communication foundation that empowers self-driving cars to interact with one another and with city infrastructure instantly. This correspondence promotes collaborative manoeuvres, improves security, and guarantees smooth navigation through intricate metropolitan surroundings.

Creating Sustainable and Efficient Urban Transportation Networks

The amalgamation of electric and self-driving vehicles with 5G technology leads to the establishment of eco-friendly and effective urban transportation networks.

Diminished Emissions: Electric automobiles, when combined with 5G-advanced charging infrastructure, diminish urban emissions and reliance on fossil fuels, contributing to fresher air and enhanced urban living circumstances.

Improved Safety: Self-driving cars, linked *via* 5G networks, share up-to-t-e-minute data regarding street circumstances, traffic movement, and possible dangers. This correspondence amplifies road safety, diminishes mishaps, and advocates for effective traffic control (Guevara & Auat Cheein, 2020).

Traffic Enhancement: Through the implementation of 5G-enabled communication, self-driving cars have the ability to foresee traffic trends, adapt routes instantly, and alleviate gridlock, resulting in seamless traffic movement and diminished travel durations.

Inclusivity: Electric and self-driving cars, combined with carpooling and mobility-as-a-service systems, enhance inclusivity in transportation, benefiting individuals without private vehicles and enhancing fairness in urban mobility.

In summary, the amalgamation of electric mobility, self-driving cars, and 5G technology offers a revolutionary chance for cities to envision urban transportation anew. By advocating for sustainability through electric automobiles and harnessing the communication capabilities of 5G for self-driving navigation, cities can establish transportation networks that are both effective and eco-friendly. This amalgamation possesses the capability to mitigate urban obstacles and pave the path for a more sustainable, interconnected, and dynamic urban future.

To examine the impact of 5G on transportation systems, consider the study by Yan *et al.*, 2020 [33]. They evaluate an evaluation system based on the self-organizing system framework of smart cities, with a focus on smart transportation systems in China.

5G FOR INTELLIGENT TRANSPORTATION SYSTEMS

The quantity of automobiles on the streets has been consistently increasing. It is anticipated that they will exceed two billion by 2030. This is owed, in part, to worldwide urbanisation, where the United Nations approximates that 21 percent of the global populace will reside in metropolises by 2050, increasing from 12 percent in 2013. This signifies grave issues to be confronted, such as the escalating count of victims by road accidents, and the degraded ecological surroundings globally. For this rationale, the notion of Smart Transportation Systems (STS) is essential to resolve the issues of transportation using Information and Communication Technology (ICT), as mentioned, to give precedence to applications that have the capacity to enhance safety, energy efficiency, traffic effectiveness, and travel convenience.

[34] present a thorough survey of 5G networks, exploring both the challenges and opportunities they present. The study discusses key technological advancements, deployment strategies, and the potential impact of 5G on various industries.

Vehicular Communication

Currently, a contemporary automobile is a sensory platform, assimilating data from the surroundings. This data is processed by an onboard processor and then utilised to aid in navigation, emission regulation, and traffic supervision, among other functions. Nevertheless, in order to attain swift data processing, it is necessary to possess an exceedingly potent on-board computer. This is the rationale behind the exorbitant price of lavish automobiles with chauffeur aid systems. To evade the utilisation of costly apparatus, *via* the Internet, it ought to be feasible to upload the data onto the cloud to execute an arduous processing load.

Therefore, IoT can aid in gathering supplementary data from traffic control hubs, supplementing the information already gathered by vehicles. With this proposition, the Automotive Cloud Computing paradigm is a demanding scenario to examine forthcoming 5G capabilities. Automobiles can swap data with other automobiles (referred to as V2V communication), with the roadside infrastructure, with the World Wide Web, with a pedestrian and likewise with any component within an intelligent metropolis. The expression Automobile to Everything or V2X is employed to designate all these categories of vehicular communication. Considering these situations, the primary patterns for automotive communications are outlined herein, introducing significant instances for security, movement, and convenience [34].

Autonomous Driving

There are several intermediate levels that involve the way we interact with vehicles, before thinking in fully autonomous ones. In particular, six levels of driving automation have been identified by the Society of Automotive Engineers (SAE), from no automation to full automation. Briefly,

No Automation (Level 0): Driver continuously in control.

Driver Assistance (Level 1): Minor driving task performed by the system.

Partial Automation (Level 2): The driver must monitor dynamic driving tasks.

Conditional Automation (Level 3): Driver does not need to monitor driving tasks, but must be able to resume control.

High Automation (Level 4): Driver not required during defined use case.

Full Automation (Level 5): The highest level refers to a fully autonomous system, with no driver required.

In essence, self-driving is feasible devoid of V2X communication, for stages 1 and 2, wherein a human operator oversees the driving surroundings. Nevertheless, stage 5, devoid of any assistance from wireless communication systems, specifically relying solely on on-board processing and sensor systems, is impracticable as it lacks automotive communication. Ambiguity must be taken into account as it cannot be guaranteed what another vehicle, or pedestrian, will execute in the upcoming moments. If automobiles distribute their accessible data, then other vehicles could utilise them to diminish ambiguities. Therefore, self-driving can be advantaged by regional communication V2V, and respond more swiftly to manoeuvres, averting crashes.

Zanella *et al.*, 2014 [35] present an extensive review of the Internet of Things (IoT), Internet of Everything (IoE), and Internet of Nano Things (IoNT) in the context of smart cities. The study explores how these IoT ecosystems contribute to the development of intelligent urban environments, highlighting their potential benefits and challenges.

A notable utilisation of self-driving technology is Collaborative Crash Prevention. As per the European Union Community Road Mishap Database (CARE), intersection-linked casualties constituted over 20% throughout the previous decade. Once all alternative traffic regulation mechanisms have faltered, the interaction among self-driving vehicles is necessary to execute measures and avert crashes. In such a vibrant and intricate setting, upon recognition of a crash hazard, vehicles cannot be determined individually, as diverse individual actions implemented without prior coordination might lead to extra crashes or unregulated circumstances. Hence, all participating vehicles ought to collaborate harmoniously to calculate the ideal collision prevention measures.

Tele-Operated Driving

It is anticipated that self-driving vehicles, in conjunction with possessing self-governing and aided driving modes, will possess the capacity to be managed by an external operator in a remote-controlled mode. This is advantageous when the self-governing mode falters, or human aid is needed in an intricate and perilous situation. Given that the driving assignment is executed from a distance, the driver's well-being deteriorates in unsafe settings like mining or subaqueous automatons.

The primary prerequisite in the teleoperation of automobiles is the real-time visual transmission from an onboard camera of the vehicle to a distant human operator to effortlessly comprehend the possible danger of the vehicle. Subsequently, relying on the live footage, the operator shall transmit movement directives to the automobile. A pivotal role is played by communication networks. Committed brief-span communications and apparent light communication are wireless communication technologies that are presently utilised but are not appropriate for far-reaching communication. The 5G cellular network is ideally suited for vehicle teleoperation as it enables covering greater distances with a substantial bandwidth and minimal delay for uninterrupted video streaming [36].

Road Safety

The manufacture of complete self-driving automobiles is expanding very rapidly, fueled by rivalry in the automotive sector and electro-mobility. Nevertheless, an elevated degree of mechanisation is not indispensable in numerous conceivable implementations. The data provided by the on-board sensors such as cameras or lasers, *via* cloud connectivity also enables the provision of cautionary applications to the drivers. In this setting, automobiles are anticipated to utilise regional V2V and worldwide V2X communication to enable a secure, further streamlined, and increasingly pleasant driving experience, as the vehicle will have the capacity to identify perilous circumstances beforehand, even if they were beyond visual range because of a curve or other automobiles in front. This will be feasible by exchanging knowledge with the neighbouring vehicles using V2V communication. In alternative terms, it would amplify the understanding that the driver possesses regarding what is transpiring around the vehicle, on the roadway. Let us contemplate the subsequent instance exhibited: a conveyance is progressing behind another one and unexpectedly, a walker is traversing the street in front of the initial automobile. The initial automobile camera identifies the circumstance and shares the picture of the pedestrian with the vehicle following it. The automobile processes the data and displays a visual warning on the windscreen along with the depiction of the pedestrian in enhanced reality. This use scenario necessitates an exceedingly elevated dependability, accessibility, minimal delay, and an elevated data speed. Some iconic road safety and traffic efficacy use cases encompass: junction collision peril alerts, urgent vehicle nearing, lane alteration caution, obscured area caution, road maintenance alerts, and road peril alerts, among other instances.

In an intelligent metropolis, security on the thoroughfare is not solely aimed at automobiles. In accordance with the National Highway Traffic Safety Administration (NHTSA), walkers make up 14% of US road deaths with more than 4400 yearly deaths. In this setting, the incorporation of Susceptible Road Users (SRUs) is going to be one of the primary objectives. The latter encompasses a broad spectrum of road users like walkers, bikers, and motorised two-wheelers. Recent investigations addressing VRU safety, for instance, primarily concentrated on identifying pedestrians and escaping mishaps utilising vehicular cloud computing as a subdivision of mobile cloud computing [36] provide a forward-looking perspective on transportation disruptions. It explores the potential collapse of the internal combustion vehicle and oil industries, emphasizing the transformative effects of new mobility technologies.

Intelligent Navigation

Self-driving cars will utilise electronic maps and geolocation to offer direction assistance to operators. These guidance services enhance driving effectiveness by selecting suitable routes based on internet traffic data. This knowledge is calculated from information provided by automobiles in the vicinity, road infrastructure or traffic control centres. Additional valuable information will be gathered with the 5G due to the Internet of Things and large-scale data, enabling the delivery of more enhanced services, and supplementing the navigation as demonstrated. A motorist will obtain notifications with customised details about points of interest (*e.g.*, sightseeing spots, eateries, parking spots, and fuel stations, among others). A synonym for "A" is "One." Analogous service is already accomplished by various mobile applications such as Google Maps, Waze and Maps.

File and Media Downloading

For the upcoming years, society will require an elevated data rate and minimal latency in the connectivity at every location and every moment. Furthermore, the implementation of autonomous automobiles could enhance the utilisation of data traffic while in motion, since operators will no longer be required to maintain vigilance while operating the vehicle and can divert their concentration towards alternative, more pleasant pursuits. This entails the augmentation in calibre and magnitude of fun activities such as internet surfing, document transfer/propagation, electronic mail, and communal platforms. Furthermore, a robust surge is anticipated in ultra-high definition footage and immersive multimedia encounters fueled by emerging technologies such as 3D footage, virtual reality, and augmented reality. This signifies a noteworthy obstacle because of the meagre network infrastructure currently implemented in highway scenarios and the rapid speed of vehicles. Capacity prerequisites, particularly because of the utilisation of high-resolution video by numerous commuters in every automobile, will generate a noteworthy burden on the cellular network. In this circumstance, effective acquisition of multimedia to travellers could be a crucial promotional tactic for self-driving. Evolution of 5G technologies, network architecture, and machine learning approaches, making it a valuable resource for understanding the state of 5G research.

ENERGY MANAGEMENT AND SUSTAINABILITY

The incorporation of fifth-generation (5G) technology within the structure of intelligent cities expands beyond urban mobility, encompassing energy administration and durability as well shown in Fig. (**3**). With the urgent requirement to restrain energy usage, diminish carbon emissions, and shift towards sustainable sources, 5G arises as a potent facilitator of energy-frugal practises and the achievement of eco-friendly urban settings. By enabling instantaneous data gathering, examination, and flexible command of energy systems, 5G empowers cities to optimise energy utilisation, integrate sustainable energy sources, and diminish their ecological impact. This amalgamation nurtures a comprehensive approach to energy administration, harmonising technological progress with the necessity of ecological guardianship to fabricate resilient and environmentally conscious cities for current and forthcoming progenies.

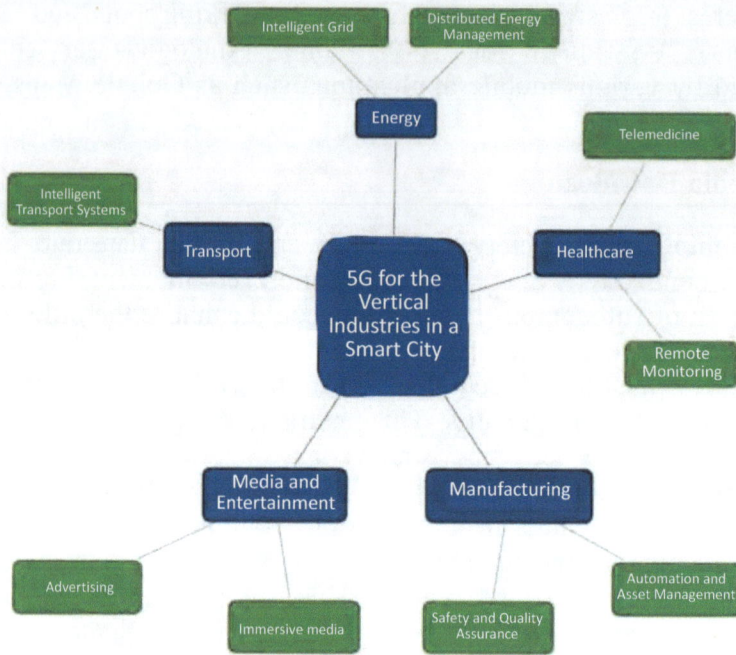

Fig. (3). Smart city vertical industries and their use cases.

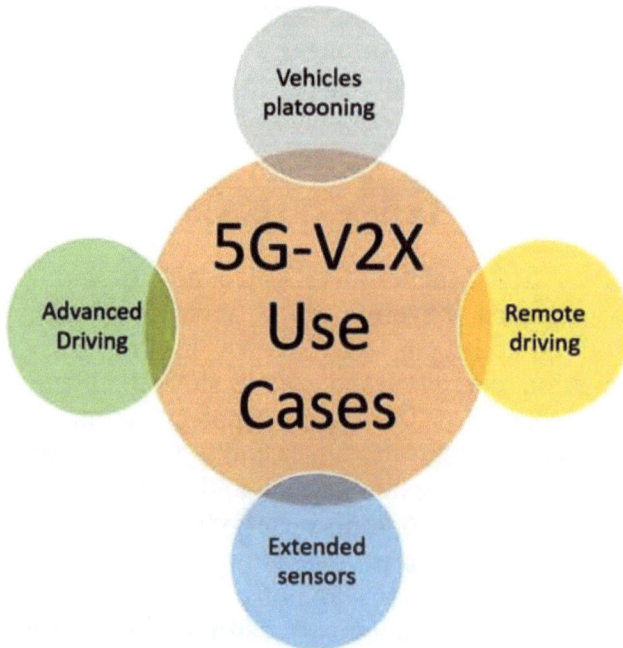

Fig. (4). Enhanced V2X use cases based on different vehicular automation levels.

Smart Energy Grids

Energy administration stands as a foundation in the quest for eco-friendly advancement within intelligent urban areas, and fifth-generation (5G) technology is a crucial instrument in this undertaking. With the burgeoning necessity to tackle energy usage and its ecological influence, the notion of intelligent energy networks arises as a revolutionary resolution.

Importance of Energy Management in Smart Cities: Efficient energy control is vital for intelligent urban areas aiming to maximise resource usage and diminish their environmental impact. Conventional energy networks frequently come up short in addressing dynamic energy requirements and integrating sustainable energy sources effectively. Intelligent energy networks tackle these deficiencies by utilising technology, up-to-the-minute information, and communication enabled by 5G to guarantee energy distribution corresponds with demand, reduces inefficiency, and promotes eco-friendly expansion.

Integration of Renewable Energy Sources and Distributed Energy Resources: Intelligent energy networks prioritise the incorporation of sustainable energy sources such as photovoltaic, breezy, and hydroelectric power. These sources provide eco-friendly and renewable alternatives to fossil fuels, contributing to diminished greenhouse gas emissions and decreased dependence on finite resources. Furthermore, the incorporation of dispersed energy assets, such as localised photovoltaic panels and power storage units, enables individual structures and neighbourhoods to produce and oversee their power, promoting self-reliance and network durability.

5G-Enabled Smart Grid Communication for Efficient Energy Distribution

The amalgamation of 5G technology and intelligent energy grids transforms energy dissemination and utilisation patterns.

Real-Time Data Analytics: 5G's elevated data velocities and reduced latency facilitate the instantaneous gathering and examination of energy usage data. This information-based approach enables cities to make knowledgeable choices and foresee energy requirements, optimising distribution for effectiveness.

Energetic Load Equilibrating: Intelligent power networks aided by 5G can flexibly adapt energy allocation according to live demand patterns. This inhibits excessive burden, diminishes energy squandering, and amplifies grid steadiness.

Demand Reaction: 5G-facilitated communication enables demand reaction initiatives, enabling energy suppliers to interact with consumers and modify energy consumption during peak periods. This reduces stress on the power network and promotes non-peak energy usage.

Microgrid Supervision: 5G innovation facilitates the transmission and synchronisation of microgrids, which are localised power networks capable of functioning autonomously or in collaboration with the primary grid. Microgrids amplify energy dependability and robustness, especially amidst grid disturbances.

In summary, the harmonisation between 5G technology and intelligent energy grids epitomises the capacity for technology to propel eco-friendly urban advancement. By incorporating sustainable energy sources, facilitating live data analysis, and promoting flexible energy distribution, intelligent cities can accomplish effective energy administration while lessening their ecological footprint. This amalgamation strengthens the dedication of astute metropolises to a forthcoming distinguished by clever asset allocation and enduring ecological viability.

Energy Consumption Monitoring

The incorporation of fifth-generation (5G) technology within intelligent cities expands its revolutionary impact on energy consumption surveillance, providing unparalleled capabilities for instantaneous examination and enhancement of energy utilisation.

Real-Time Monitoring and Analysis of Energy Usage: Efficient energy administration relies on precise and prompt observations of energy usage trends. 5G technology enables the instantaneous surveillance and examination of energy consumption throughout different urban domains. By furnishing elevated data velocities and diminished latency, 5G empowers uninterrupted data compilation, guaranteeing that urban authorities and consumers alike possess current information regarding energy requisition, dissemination, and squandering.

5G-Enabled Smart Meters and Sensors for Data Collection

Intelligent gauges and detectors empowered by 5G are essential components of energy usage monitoring.

Intelligent Gauges: 5G-powered intelligent gauges substitute conventional utility gauges, enabling the distant and instantaneous quantification of electricity, water, and gas consumption. These gauges provide precise information that enlightens invoicing, usage trends, and energy-conserving tactics.

Detectors and Internet of Things Gadgets: 5G facilitates the implementation of detectors and Internet of Things (IoT) gadgets across the metropolitan scenery.

These detectors gather information on illumination, warming, chilling, and appliance utilisation, contributing to a comprehensive comprehension of energy expenditure patterns.

Enabling Demand-Response Strategies for Energy Conservation

One of the crucial advantages of 5G-facilitated energy consumption monitoring is the capability to execute demand-response tactics.

Live Feedback: With immediate access to energy usage data, consumers receive live feedback on their energy consumption behaviours. This feedback motivates energy-conscious choices and actions that contribute to preservation.

Demand-Reply Initiatives: 5G technology enables demand-response initiatives in which energy providers communicate directly with consumers to modify energy usage during peak demand periods. By willingly decreasing energy usage when the grid is strained, consumers contribute to grid stability and preservation endeavours.

Peak Load Management: By observing energy usage patterns in real-time, cities can foresee peak demand periods and execute load management tactics that diminish energy consumption during these crucial times [37, 38].

In summary, 5G-facilitated energy usage surveillance highlights the possibility for technology to propel energy effectiveness and preservation in intelligent urban areas. By furnishing instantaneous perceptions through intelligent gauges, detectors, and Internet of Things (IoT) gadgets, 5G enables consumers, municipal authorities, and energy suppliers to enhance energy utilisation, execute demand-response tactics, and diminish the overall ecological influence of urban existence. This amalgamation signifies a noteworthy stride towards attaining enduring energy practises within the framework of intelligent urban areas.

5GAmericas, (2019) offers a comprehensive overview of 5G services and use cases. The study discusses various applications of 5G technology, including enhanced mobile broadband, IoT, and mission-critical communications.

To gain insights into the standardization and architectural aspects of 5G, Agiwal *et al.*, (2016) offer an in-depth review. The study covers the standardization bodies involved in 5G development and discusses the fundamental architecture of 5G networks.

ENVIRONMENTAL MONITORING AND WASTE MANAGEMENT

The incorporation of fifth-generation (5G) technology within intelligent metropolis frameworks expands its revolutionary impact on ecological surveillance and garbage administration, nurturing a more eco-friendly urban milieu. With metropolitan areas grappling with predicaments such as contamination and insufficient waste disposal practises, the convergence of 5G and intelligent city initiatives offers a resilient solution. Leveraging 5G-equipped sensors, urban areas can collect up-to-the-minute information on atmosphere and liquid quality, sound levels, and other crucial ecological markers, enabling proactive reactions to emerging issues. Furthermore, 5G-fueled Internet of Things (IoT) gadgets transform waste management by enhancing collection routes *via* up-to-the-minute fill-level information from intelligent waste containers. This approach improves waste collection effectiveness, diminishes superfluous pickups, and empowers data-based waste diversion tactics for heightened recycling. As cities harness 5G technology for ecological surveillance and refuse administration, they enable themselves to establish purer, fitter, and increasingly sustainable urban panoramas for present and forthcoming progenies.

Air Quality and Environmental Monitoring

Incorporating fifth-gen (5G) technology into intelligent metropolis frameworks presents a groundbreaking method to oversee air and water purity, greatly amplifying ecological administration tactics and overall urban welfare. The function of 5G in this context is crucial, as it offers the necessary communication infrastructure to facilitate instantaneous data gathering, examination, and distribution.

The implementation of 5G-equipped sensors establishes the foundation of efficient ecological surveillance. These detectors are strategically placed throughout metropolitan regions to consistently seize information concerning contaminants, fine particles, fumes, and other ecological markers. Harnessing 5G's elevated data velocities and minimal latency, these sensors transmit data to central platforms without any postponement, guaranteeing that decision-makers have access to precise and current information.

The live data gathered by these sensors undergoes sophisticated analysis, producing valuable understandings of contamination levels, patterns, and possible health consequences. This perceptive capacity empowers cities to adopt proactive measures by recognising pollution origins, evaluating the efficiency of ecological strategies, and making enlightened choices that prioritise public well-being and ecological durability.

Moreover, the incorporation of 5G and ecological surveillance enables premature identification of contamination and ecological perils. By acquiring instant data notifications from sensors indicating surges in contamination levels, cities can promptly react with suitable interventions. This punctual approach not only protects the welfare of inhabitants but also enables cities to tackle pollution sources promptly, averting prolonged ecological deterioration.

Guevara & Auat Cheein, (2020) explore the pivotal role of 5G technologies in smart cities and intelligent transportation systems. The study discusses the challenges and prospects of integrating 5G to enhance urban mobility and connectivity.

In essence, the harmonisation between 5G technology and ecological surveillance is a pivotal stride towards establishing purer, fitter, and further sustainable urban surroundings. By offering instantaneous observations, facilitating information-based decision-making, and enabling prompt reactions to emerging issues, 5G empowers intelligent cities to proactively govern their environment, guaranteeing an enhanced standard of living for inhabitants while reducing the environmental impact of urban existence.

Role of 5G in Monitoring Air and Water Quality: The commencement of fifth-generation (5G) technology transforms the surveillance of atmospheric and aquatic conditions within intelligent urban areas. By offering super-fast data speeds and minimal latency, 5G enables the smooth transmission of real-time ecological data gathered by sensors. This capacity enables cities to oversee air and water quality with unparalleled precision and promptness, enabling knowledgeable decision-making, timely interventions, and proactive administration of environmental issues.

Deploying Sensors for Real-Time Data Collection and Analysis: The commencement of fifth-generation (5G) technology transforms the surveillance of atmospheric and aquatic conditions within intelligent urban areas. By offering super-fast data speeds and minimal latency, 5G enables the smooth transmission of real-time ecological data gathered by sensors. This capacity enables cities to oversee air and water quality with unparalleled precision and promptness, enabling knowledgeable decision-making, timely interventions, and proactive administration of environmental issues.

Early Detection of Pollution and Environmental Hazards: The collaboration of 5G technology and ecological surveillance enables cities to identify contamination and ecological risks in their early phases. By acquiring immediate notifications from sensors indicating surges in contamination levels, urban authorities can react swiftly with suitable actions. This proactive strategy not only protects public health but also enables the detection of contamination origins. Premature identification empowers urban areas to take prompt measures, averting additional ecological deterioration and minimising the enduring consequences of contamination.

In summary, the incorporation of 5G technology into ecological surveillance amplifies the effectiveness, precision, and promptness of metropolitan ecological administration. By enabling instantaneous data gathering, examination, and premature identification, 5G empowers intelligent metropolises to address contamination, guarantee a more salubrious habitation milieu, and foster enduring viability.

Smart Waste Management

Intelligent waste management signifies a noteworthy progression within the domain of intelligent urban areas, with next-generation (5G) technology playing a pivotal role in enhancing waste elimination methods, lessening ecological influence, and fostering effective resource utilisation.

Incorporation of 5G for Real-Time Data Compilation: 5G innovation facilitates the deployment of Internet of Things (IoT) devices and sensors that monitor waste receptacles' capability in real-time. These 5G-enabled sensors transmit data seamlessly, allowing waste management systems to collect accurate information about waste generation patterns and optimize collection routes accordingly.

Dynamic Waste Collection and Efficiency: With 5G-equipped sensors, waste management systems can flexibly adapt collection timetables based on fill-level information. This live-time optimisation minimises unnecessary waste collection journeys, decreases fuel usage, and diminishes greenhouse gas discharges. The effective waste gathering procedure additionally contributes to tidier metropolitan settings and improved visual appeal [36].

Data-Driven Waste Diversion Strategies: 5G-fueled intelligent refuse management systems not only optimise waste collection logistics but also empower data-driven waste diversion strategies. By examining waste composition and generation patterns, cities can prioritise recycling initiatives, promote appropriate waste disposal behaviours, and reduce the quantity of waste dispatched to landfills.

In essence, the incorporation of 5G technology into intelligent waste management revolutionises conventional waste elimination practises into effective, information-based resolutions. This amalgamation contributes to the establishment of greener, more sustainable cities by diminishing waste-related ecological consequences and augmenting the overall calibre of urban living.

Obstacles in Urban Waste Management: Urban waste management presents notable hurdles, comprising ineffective gathering techniques, insufficient disposal infrastructure, and ecological worries linked to waste buildup. Conventional waste management systems frequently face challenges in maximising collection routes, resulting in heightened fuel usage, discharges, and operational expenses. In this setting, the incorporation of fifth-generation (5G) technology presents inventive resolutions that revolutionise waste management procedures within intelligent cities.

5G-Powered Solutions for Intelligent Garbage Receptacles and Assortment: 5G technology transforms waste management by facilitating the implementation of intelligent waste containers furnished with detectors and Internet of Things (IoT) gadgets. These 5G-equipped sensors observe fill levels in real time, transmitting data effortlessly to waste management systems. This live data gathering enables cities to accurately monitor waste production patterns, guaranteeing effective waste collection tactics and prompt bin clearing.

Maximisation of Waste Collection Routes for Decreased Emissions: The collaboration of 5G technology and intelligent waste bins enables cities to enhance waste collection routes dynamically. By scrutinising live fill level data and capitalising on 5G's elevated data velocities, garbage management systems can adapt collection timetables in reaction to shifting waste production patterns. This optimisation diminishes superfluous journeys, diminishes fuel usage, diminishes greenhouse gas discharges, and improves overall operational effectiveness.

In summary, the amalgamation of 5G technology and intelligent waste management tackles urban waste predicaments by providing effective, information-based resolutions. The implementation of 5G-equipped detectors and flexible gathering techniques enhances waste collection pathways, resulting in diminished ecological influence, enhanced resource utilisation, and tidier metropolitan surroundings. This amalgamation demonstrates the dedication of intelligent urban areas to improve waste management strategies for the advantage of both inhabitants and the globe.

CITIZEN ENGAGEMENT AND QUALITY OF LIFE

Fifth-gen (5G) technology possesses the capability to revolutionise the connection between individuals and their metropolitan surroundings in the framework of intelligent cities, introducing a fresh epoch of citizen involvement and enhanced standard of living. By providing rapid connectivity, minimal delay, and improved data interchange capabilities, 5G enables smooth interactions between citizens and city services. This facilitates the establishment of dynamic platforms for citizen engagement, input gathering, and live communication with municipal authorities. By virtue of 5G-fueled applications like cellular applications, intelligent gadgets, and enhanced reality encounters, individuals can acquire knowledge about communal amenities, conveyance, ecological circumstances, and artistic occasions in a moment. This increased connectivity promotes a feeling of communal possession, empowering individuals to actively participate in decision-making procedures, tackle urban obstacles, and collectively mould the destiny of their cities. Ultimately, the amalgamation of 5G technology empowers intelligent metropolises to enrich citizen involvement, cultivate inclusiveness, and elevate the overall calibre of existence for inhabitants by enabling them to participate actively in moulding the urban panorama.

Smart City Services and Citizen Engagement

The incorporation of fifth-generation (5G) technology within intelligent urban services presents a revolutionary pathway for improving citizen involvement and overall standard of living. By harnessing the potentials of 5G, cities can provide inventive and adaptable services that empower citizens to actively engage in moulding their urban surroundings.

Engaging and Live Services: 5G technology's elevated data velocities and reduced latency facilitate the provision of interactive and instantaneous intelligent urban services directly to citizens' devices. By means of mobile applications, online portals, and digital platforms, individuals can obtain data, offer input, and interact with civic amenities effortlessly. This connectivity enables immediate access to up-to-the-minute information on transportation, communal amenities, ecological circumstances, and beyond, augmenting citizens' capacity to formulate well-informed choices [38].

Customised Experiences and Enhanced Reality: The fusion of 5G and augmented reality (AR) technologies generates tailored and captivating encounters for individuals. With AR-capable gadgets, inhabitants can traverse urban areas using improved navigation tools, obtain historical facts about landmarks, and participate in cultural occasions through virtual displays. This interactive stratum enhances citizens' interactions with their surroundings, cultivating a more profound sense of connection and involvement.

Input from citizens and stakeholders is crucial for effective decision-making and the establishment of feedback loops. The integration of 5G technology in citizen engagement platforms facilitates the creation of efficient feedback loops and promotes collaborative decision-making. Inhabitants can express worries, propose ideas, and engage in public discussions in live time, enabling local authorities to react promptly to urban difficulties and integrate resident viewpoints into decision-making procedures. This openness and engagement augment citizen confidence and possession of urban initiatives.

In summary, the incorporation of 5G technology into intelligent urban services enhances citizen involvement to unprecedented levels. By providing engaging, live experiences, customised engagements, and channels for cooperative decision-making, cities enable citizens to engage in an energetic role in shaping the urban scenery, nurturing a more comprehensive, participative, and dynamic standard of living for all inhabitants.

Enhancing Public Services *via* Digital Platforms: The arrival of fifth-generation (5G) technology introduces a fresh epoch of improving citizen services through inventive digital platforms within intelligent cities. Harnessing 5G's potential, cities can transform the way citizens engage with municipal services, cultivating a more effective and user-focused urban encounter.

5G-Advanced Communication for Citizen-Government Engagement: 5G's elevated data speeds and minimal latency establish the groundwork for smooth and immediate communication between citizens and municipal authorities. This facilitates instantaneous interactions *via* digital platforms, enabling citizens to participate with government agencies, inquire for information, and offer feedback without obstacles. The improved communication enables prompt problem-solving, diminishes administrative obstacles, and fosters increased open citizen-government connections.

Tailored Services and Instantaneous Information Dissemination: The incorporation of 5G technology empowers cities to provide individualised and situation-sensitive services to residents. By means of data analysis and instantaneous information distribution, individuals can obtain personalised suggestions, notifications, and revisions concerning transportation, civic amenities, occasions, and additional matters. This customised approach not only enhances citizens' convenience but also enables them to make well-informed choices about their daily endeavours.

To contextualize the emergence of smart cities, (United Nations, 2018; United Nations, 2019) provide a global perspective on urbanization trends and smart city initiatives. These studies offer valuable data and insights into the rapid urbanization of regions worldwide and the corresponding efforts to create sustainable and technology-driven urban environments.

(Attaran, 2021) provides an extensive review of the impact of 5G on industry digitization. It delves into how 5G technologies are reshaping industrial processes, automation, and intelligent manufacturing, offering insights into the evolving landscape of Industry 4.0.

In essence, 5G technology's incorporation into citizen services through digital platforms paves the way for a more vibrant, reactive, and user-centric urban environment. By offering smooth communication, individualised experiences, and immediate access to information, intelligent cities enhance the calibre of citizen services, promoting a more robust feeling of community involvement and contentment among inhabitants.

Enhancing Quality of Life

The incorporation of fifth-generation (5G) technology into intelligent city initiatives holds the potential of improving the standard of living for inhabitants by reimagining how they engage with their metropolitan surroundings. By means of the capacities of 5G, cities can introduce a plethora of innovations that contribute to enhanced welfare and a more gratifying urban experience.

Flawless Connectivity and Availability: 5G's rapid-speed connectivity and minimal delay enable smooth access to a multitude of services and assets, establishing a linked environment that streamlines everyday life. Inhabitants can participate in distant labour, internet-based learning, telemedicine services, and virtual social engagements with minimal interruptions, resulting in heightened convenience and efficiency.

Intelligent Solutions for Metropolitan Convenience: The incorporation of 5G-fueled Internet of Things (IoT) gadgets enables the execution of intelligent services that amplify urban convenience. Intelligent illumination, weather regulation, and effective energy administration systems contribute to a more enjoyable and energy-saving living atmosphere, guaranteeing that inhabitants experience supreme satisfaction while reducing ecological influence.

Cultural Enhancement and Enhanced Reality: By merging 5G technology with augmented reality (AR), cities can provide immersive cultural encounters that enhance residents' lives. Enhanced reality excursions, interactive past locations, and virtual artistic showcases contribute to a profound affiliation with the city's legacy and traditions, nurturing a feeling of belonging and admiration among inhabitants.

Utilising 5G for Healthcare, Education, and Entertainment: The incorporation of fifth-generation (5G) technology within intelligent city frameworks unveils fresh possibilities for revolutionising healthcare, education, and entertainment, restructuring the urban encounter and promoting an enhanced standard of living for inhabitants.

Distant Healthcare Services, Cyber Learning, and Cultural Encounters: 5G's elevated data velocities and minimal latency empower a plethora of revolutionary applications.

Remote Healthcare Services: 5G enables smooth remote healthcare services, enabling residents to access virtual consultations, track health metrics, and receive medical advice from the convenience of their homes. This improves healthcare availability, especially for individuals with restricted mobility or those residing in distant regions.

Virtual Learning: The amalgamation of 5G technology sustains virtual learning platforms that provide interactive and captivating educational experiences. Learners can avail online courses, join in cooperative assignments, and interact with educational material that surpasses the confines of a traditional classroom.

Cultural Encounters: 5G-fueled enhanced and simulated reality applications offer inhabitants with enhanced cultural encounters. By means of AR-assisted urban excursions, virtual historical locations, and interactive artistic exhibits, inhabitants can actively participate in their city's legacy and cultural provisions in inventive and captivating manners.

Promoting a Higher Quality of Life for Residents

The implementation of 5G technology throughout healthcare, education, and entertainment domains collectively advances a superior standard of living for inhabitants.

Fitness and Wellness: 5G-facilitated distant healthcare services empower residents to access medical aid conveniently, promoting early identification, precautionary care, and well-being.

Education Accessibility: Virtual instruction *via* 5G-boosted platforms guarantees educational persistence, regardless of bodily constraints, promoting uninterrupted learning possibilities and educational fairness.

Cultural Involvement: Inhabitants' involvement with their urban area's cultural legacy is intensified through 5G-boosted cultural encounters, nurturing a feeling of satisfaction, affiliation, and community linkage.

SUMMARY

In the ever-changing terrain of intelligent metropolises, the incorporation of fifth-generation (5G) technology arises as a revolutionary power propelling urban advancement towards durability, heightened citizen involvement, and upgraded standard of living. The chapter explores the complex interplay between 5G technology, intelligent city initiatives, eco-friendly technology, and the progress of urban living standards.

The foundation is established by investigating the fundamental ideas of intelligent cities, highlighting their central elements such as technology incorporation, data utilisation, and a concentration on sustainability. The mutualistic connection between eco-friendly technology and intelligent metropolises is revealed, displaying how they seamlessly contribute to ecological conservation, asset effectiveness, and a more sustainable urban future.

The limelight then moves to 5G technology, unveiling its essential characteristics such as elevated data speeds, reduced latency, and extensive device connectivity. The amalgamation of 5G with the Internet of Things (IoT) sparks a revolution in urban mobility, intelligent transportation systems, and the progression of electric and self-driving vehicles. The technology's function in energy administration, sustainability, and ecological surveillance emphasizes its importance in advancing efficient energy utilisation, instantaneous environmental information gathering, and successful waste control methods.

Moreover, the chapter highlights the metamorphic potential of 5G in augmenting citizen participation and standard of living. By facilitating smooth connectivity, customised services, and engaging experiences, 5G empowers individuals to actively participate in moulding their urban surroundings. The amalgamation of 5G technology with healthcare, education, and entertainment magnifies distant healthcare services, virtual learning, and cultural experiences, nurturing an enhanced quality of life for all inhabitants.

REFERENCES

[1]   3GPP, "Technical Specification Group Services and System Aspects; System Architecture for the 5G System (5GS); Stage 2 (Release 16)," 2020. Available from: https://portal.3gpp.org/desktopmodules/ Specifications/SpecificationDetails.aspx?specificationId=3144 [Accessed: May 3, 2021].

[2]   5G Americas, "5G Services and Use Cases," 2019. Available from: https://www.5gamericas.org/5g-services-use-cases/ [Accessed: May 3, 2021].

[3]   S.A. Abdel Hakeem, A.A. Hady, and H. Kim, "5G-V2X: standardization, architecture, use cases, network-slicing, and edge-computing", *Wirel. Netw.*, vol. 26, no. 8, pp. 6015-6041, 2020. [http://dx.doi.org/10.1007/s11276-020-02419-8]

[4]   Accenture Strategy, "The Impact of 5G on the European Economy," 2021. Available from: https://www.accenture.com/_acnmedia/PDF-144/Accenture-5G-WP-EU-Feb26.pdf [Accessed: May 3,

2021].

[5]   M. Agiwal, A. Roy, and N. Saxena, "Next Generation 5G Wireless Networks: A Comprehensive Survey", *IEEE Commun. Surv. Tutor.,* vol. 18, no. 3, pp. 1617-1655, 2016.
[http://dx.doi.org/10.1109/COMST.2016.2532458]

[6]   N. Al-Falahy, and O.Y. Alani, "Technologies for 5G Networks: Challenges and Opportunities", *IT Prof.,* vol. 19, no. 1, pp. 12-20, 2017.
[http://dx.doi.org/10.1109/MITP.2017.9]

[7]   J. Arbib and T. Seba, "Rethinking Transportation 2020–2030—The Disruption of Transportation and the Collapse of the Internal-Combustion Vehicle and Oil Industries," 2017. Available from: https://static1.squarespace.com/static/585c3439be65942f022bbf9b/t/591a2e4be6f2e1c13df930c5/1509 063152647/RethinkX%2bReport_051517.pdf [Accessed: May 3, 2021].
[http://dx.doi.org/10.61322/XWUI2081]

[8]   M. Attaran, "The impact of 5G on the evolution of intelligent automation and industry digitization", *J. Ambient Intell. Humaniz. Comput.,* vol. 12, pp. 1545-1554, 2021.
[PMID: 33643481]

[9]   H. Bagheri, M. Noor-A-Rahim, Z. Liu, H. Lee, D. Pesch, K. Moessner, and P. Xiao, "5G NR-V2X: Toward Connected and Cooperative Autonomous Driving", *IEEE Communications Standards Magazine,* vol. 5, no. 1, pp. 48-54, 2021.
[http://dx.doi.org/10.1109/MCOMSTD.001.2000069]

[10]   G. Brown, "Private 5G Mobile Networks for Industrial IoT," *Qualcomm,* 2019. Available from: https://www.qualcomm.com/media/documents/files/private-5g-networks-for-industrial-iot.pdf [Accessed: May 3, 2021].

[11]   Digital 2020, "Global Digital Overview," 2020. Available from: https://datareportal.com/reports/ digital-2020-global-digital-overview [Accessed: May 3, 2021].

[12]   E. Fadel, V.C. Gungor, L. Nassef, N. Akkari, M.G.A. Malik, S. Almasri, and I.F. Akyildiz, "A survey on wireless sensor networks for smart grid", *Comput. Commun.,* vol. 71, pp. 22-33, 2015.
[http://dx.doi.org/10.1016/j.comcom.2015.09.006]

[13]   H. Fourati, R. Maaloul, and L. Chaari, "A survey of 5G network systems: challenges and machine learning approaches", *Int. J. Mach. Learn. Cybern.,* vol. 12, no. 2, pp. 385-431, 2021.
[http://dx.doi.org/10.1007/s13042-020-01178-4]

[14]   L. Guevara, and F. Auat Cheein, "The Role of 5G Technologies: Challenges in Smart Cities and Intelligent Transportation Systems", *Sustainability (Basel),* vol. 12, no. 16, p. 6469, 2020.
[http://dx.doi.org/10.3390/su12166469]

[15]   ] Infotech, "Utilities in Europe Accelerate Digital Transformation," 2020. Available from: https://infotechlead.com/cio/utilities-in-europe-acceleratedigital-transformation-62370 [Accessed: May 3, 2021].

[16]   IoT-Analytics, "The Top 10 IoT Segments in 2018—Based on 1600 Real IoT Projects," 2018. Available from: https://iot-analytics.com/top-10-iot-segments-2018-real-iot-projects [Accessed: May 3, 2021].

[17]   ITU, "IMT Vision—Framework and Overall Objectives of the Future Development of IMT for 2020 and beyond," 2015. Available from: https://www.itu.int/rec/R-REC-M.2083 [Accessed: May 3, 2021].

[18]   S. Jewkes and C. Steitz, "Power to the Drones: Utilities Place Bets on Robots," 2018. Available from: https://www.reuters.com/article/ctech-us-utilities-drones-europe-analysi-idCAKBN1K60TS-OCATC [Accessed: May 3, 2021].

[19]   S. Joss, F. Sengers, D. Schraven, F. Caprotti, and Y. Dayot, "The Smart City as Global Discourse: Storylines and Critical Junctures across 27 Cities", *J. Urban Technol.,* vol. 26, no. 1, pp. 3-34, 2019.
[http://dx.doi.org/10.1080/10630732.2018.1558387]

[20]   P.V. Klaine, M.A. Imran, O. Onireti, and R.D. Souza, "A Survey of Machine Learning Techniques Applied to Self-Organizing Cellular Networks", *IEEE Commun. Surv. Tutor.*, vol. 19, no. 4, pp. 2392-2431, 2017.
       [http://dx.doi.org/10.1109/COMST.2017.2727878]

[21]   K. Ma, X. Liu, Z. Liu, C. Chen, H. Liang, and X. Guan, "Cooperative Relaying Strategies for Smart Grid Communications: Bargaining Models and Solutions", *IEEE Internet Things J.*, vol. 4, no. 6, pp. 2315-2325, 2017.
       [http://dx.doi.org/10.1109/JIOT.2017.2764941]

[22]   M.H. Miraz, M. Ali, P.S. Excell, and R. Picking, "A review on Internet of Things (IoT), Internet of Everything (IoE) and Internet of Nano Things (IoNT)", *Proceedings of the 2015 Internet Technologies and Applications (ITA)*, pp. 219-224, 2015.
       [http://dx.doi.org/10.1109/ITechA.2015.7317398]

[23]   J.F. Monserrat, G. Mange, V. Braun, H. Tullberg, G. Zimmermann, and Ö. Bulakci, "METIS research advances towards the 5G mobile and wireless system definition", *EURASIP J. Wirel. Commun. Netw.*, vol. 2015, no. 1, p. 53, 2015.
       [http://dx.doi.org/10.1186/s13638-015-0302-9]

[24]   G. Nencioni, R.G. Garroppo, A.J. Gonzalez, B.E. Helvik, and G. Procissi, "Orchestration and Control in Software-Defined 5G Networks: Research Challenges", *Wirel. Commun. Mob. Comput.*, vol. 2018, no. 1, p. 6923867, 2018.
       [http://dx.doi.org/10.1155/2018/6923867]

[25]   J.T.J. Penttinen, "Positioning of 5G", In: *5G Explained: Security and Deployment of Advanced Mobile Communications.* Wiley Telecom: Hoboken, NJ, USA, 2019, pp. 47-70.
       [http://dx.doi.org/10.1002/9781119275695.ch3]

[26]   P. Popovski *et al.*, "D6.6 Final Report on the METIS 5G System Concept and Technology Roadmap," 2014. Available from: https://riunet.upv.es/handle/10251/76765 [Accessed: May 3, 2021].

[27]   J. Porter, "Go Read This Analysis of What the iPad Pro's LIDAR Sensor Is Capable of," 2020. Available from: https://www.theverge.com/2020/4/16/21223626/ipad-pro-halide-camera-lidar sensor-augmented-reality-scanning

[28]   PSB Research, "5G Economy Global Public Survey Report Commissioned by Qualcomm," 2019. Available from: https://www.qualcomm.com/media/documents/files/psb-public-survey-report.pdf [Accessed: May 3, 2021].

[29]   S.R. S, T. Dragičević, P. Siano, and S.R.S. Prabaharan, "Future generation 5g wireless networks for smart grid: A comprehensive review", *Energies,* vol. 12, no. 11, p. 2140, 2019.
       [http://dx.doi.org/10.3390/en12112140]

[30]   R. Sánchez-Corcuera, A. Nuñez-Marcos, J. Sesma-Solance, A. Bilbao-Jayo, R. Mulero, U. Zulaika, G. Azkune, and A. Almeida, "Smart cities survey: Technologies, application domains and challenges for the cities of the future", *Int. J. Distrib. Sens. Netw.*, vol. 15, no. 6, 2019.
       [http://dx.doi.org/10.1177/1550147719853984]

[31]   Y. Song, J. Lin, M. Tang, and S. Dong, "An internet of energy things based on wireless LPWAN", *Engineering (Beijing),* vol. 3, no. 4, pp. 460-466, 2017.
       [http://dx.doi.org/10.1016/J.ENG.2017.04.011]

[32]   A.S. Syed, D. Sierra-Sosa, A. Kumar, and A. Elmaghraby, "IoT in Smart Cities: A Survey of Technologies, Practices and Challenges", *Smart Cities,* vol. 4, no. 2, pp. 429-475, 2021.
       [http://dx.doi.org/10.3390/smartcities4020024]

[33]   H. K. Trabish, "Demand Response Failed California 20 Years Ago; the State's Recent Outages May Have Redeemed It," 2020. Available from: https://www.utilitydive.com/news/demand-response failed-california-20-years-ago-the-states-recent-outages/584878/ [Accessed: May 3, 2021].

[34]   United Nations, *World Urbanization Prospects: The 2018 Revision.* United Nations: New York, NY,

USA, 2019.

[35] United Nations, *World Urbanization Prospects.* United Nations: New York, NY, USA, 2018.

[36] J. Wang, H. Zhong, Q. Xia, and C. Kang, "Optimal Planning Strategy for Distributed Energy Resources Considering Structural Transmission Cost Allocation", *IEEE Trans. Smart Grid,* vol. 9, no. 5, pp. 5236-5248, 2018.
[http://dx.doi.org/10.1109/TSG.2017.2685239]

[37] L. Wood, "Global smart cities market report 2020–2025: Analysis & Forecasts of Smart Transportation, Smart Buildings, Smart Utilities, Smart Citizen Services," 2020. Available from: https://www.businesswire.com/news/home/20201008005413/en/Global-Smart-Cities-Market-Report-2020-2025-Analysis-Forecasts-of-Smart-Transportation-Smart-Buildings-Smart-Utilities-Smart-Citizen-Services—Research [Accessed: May 3, 2021].

[38] J. Yan, J. Liu, and F.M. Tseng, "An evaluation system based on the self-organizing system framework of smart cities: A case study of smart transportation systems in China", *Technol. Forecast. Soc. Change,* vol. 153, p. 119371, 2020.
[http://dx.doi.org/10.1016/j.techfore.2018.07.009]

[39] A. Zanella, N. Bui, A. Castellani, L. Vangelista, and M. Zorzi, "Internet of things for smart cities", *IEEE Internet Things J.,* vol. 1, no. 1, pp. 22-32, 2014.
[http://dx.doi.org/10.1109/JIOT.2014.2306328]

[40] Nataraju, D. Pradhan, and S. S. Jambli, "Opportunities, challenges, and benefits of 5G-IoT toward sustainable development of green smart cities (SD-GSC)," in *3rd International Conference on Intelligent Technologies (CONIT)*, 2023, pp. 1–8,

[41] D. Pradhan, R. Suma, Rajeswari, and C. B. Carva, "Implications of resource optimization in D2D communication for efficient usage of RF energy: Practice and challenges," in 2023 3rd International Conference on Intelligent Technologies (CONIT), 2023, pp. 1–8.

[42] K. Sinha, D. Pradhan, A. Saurabh, and A. Kumar, "Software Principles of 5G Coverage: Simulator Analysis of Various Parameters", In: *The Software Principles of Design for Data Modeling.* IGI Global, 2023, pp. 276-285.
[http://dx.doi.org/10.4018/978-1-6684-9809-5.ch020]

[43] H.K. Sinha, A. Kumar, and D. Pradhan, "A Study of Various Peak to Average Power Ratio (PAPR) Reduction Techniques for 5G Communication System (5G-CS)", In: *Optimization Techniques in Engineering: Advances and Applications*, 2023, pp. 239-254.
[http://dx.doi.org/10.1002/9781119906391.ch27]

[44] D. Pradhan, P.K. Sahu, H.M. Tun, and N.K. Wah, "Integration of AI/Ml in 5g technology toward intelligent connectivity, security, and challenges", In: *Machine Learning Algorithms and Applications in Engineering.* CRC Press, 2023, pp. 239-254.
[http://dx.doi.org/10.1201/9781003104858-14]

[45] P. K. Priyanka, G. Mallavaram, A. Raj, D. Pradhan, and Rajeswari, "Cognitiveness of 5G Technology toward sustainable development of smart cities," in *Decision Support Systems for Smart City Applications*, eta 2022, pp. 189–203.

CHAPTER 5

5G–Enabled Smart Healthcare System with the Integration of Blockchain Technology

Sindhu Rajendran[1,*]**, P. Kalyan Ram**[1] **and Akash Kotagi**[1]

[1] *Department Electronics and Communication Engineering, Rashtreeya Vidyalaya College of Engineering, Bengaluru, India*

Abstract: Among the most crucial jobs in the digitization era is to track the data in real-time for a wide network of healthcare systems. Blockchain technology introduces us to the new age of sharing information in an authorized way using different consensus algorithms to connect the data blocks in chains, along with the help of Hash-keys making it safer. In blockchain technology, any entrance of malicious data replacing the original data cannot be encouraged because the distributed ledgers have the same data in an encrypted manner, and changing the same data in such a huge network is merely impossible, this enhances the security of the user's information. Smart healthcare systems on a higher basis use Health Information Exchange(HIE), to decentralize the previous health records of the patient between organizations and frequently update them. Smart healthcare systems make it viable for decentralization of patient data, the number of drugs consumed, and statistics of different diseases as 5G plays a major role here because of its distributed implementation, its connectivity with IoTs and IIoTs that help in the easy update of information with the patient's access given. With the advent of 5G-NR, using the modulation techniques of QAM, variable Bandwidth, and the NOMA, it has enhanced higher data rates and high networking capacities. Mobile Edge Computing (MEC) of 5G technology helps in storing and computing data in a decentralized manner with the help of distributed Mobile Cloud Centres(MCC). Over time, many private blockchain technologies have been suggested, which involve only a few organizations and transact data only between them unlike the public blockchain technology thereby increasing the reliability and security. In this chapter, we emphasize smart health sectors, the necessity of blockchain, the different blockchain designs for healthcare applications, and the different proposed algorithms based on 5G, and the chapter concludes with recent advances in 5G networks, the challenges, and potential solutions.

Keywords: Blockchain, Cloud computing, MEC, Smart healthcare, 5G-NR.

---

[*] **Corresponding author Sindhu Rajendran:** Department Electronics and Communication Engineering, Rashtreeya Vidyalaya College of Engineering, Bengaluru, India; E-mail: sindhur@rvce.edu.in

Devasis Pradhan, Mangesh M. Ghonge, Nitin S. Goje, Alessandro Bruno and Rajeswari (Eds.)

INTRODUCTION

Evolution of Block-Chain

There has been a tremendous change in technologies within the last 30 years of period, the rise in the software industry has taken the service sector into a golden period which has a lot of things to deal with such as network chain, information, availability, security, transaction of information, statistics on purchases, *etc.* The above domains have revolutionized each and every sector such as automobile, internet, wireless communication, business, banking, health, education, *etc.* Before, several different industries carried out all of these tasks, but there were numerous flaws and challenges, including: a) disturbance and lack of integrity in the network chain because of the centralized system. b) Information should always be retrieved when needed from servers which then takes a lot of power and eats up In-memory. c) Disruption of the chains because of lack of rules and gathering all the confidential information of a person or an organization by creating breaches or corrupting them by adding viruses in the software or websites through which we cite. d) Centralized transaction of information with different rules and restrictions of different entities. e) In the field of AI, statistics is the most important for data to be asserted, even though there are a lot of information transactions made in all sectors, with the inclusion of software, there is no proper update of data from time to time to perform proper models to predict further developments for the future. There are still many problems that can be addressed like remote area access with these technologies and that is when everyone started looking for new technologies that can optimize all these problems and increase performance and the new solutions are solved using the technologies shown in Fig. (**1**). This paved the way for the significance of Block-chain that was first introduced in 2008 with an application of bitcoin by SatoshiNakamoto in his paper for building a trusted model for transactions that do not rely on financial and third party organizations with an idea of distributed ledger network with an end to end connection. With the tremendous growth of Block-chain in recent time periods, this technology has no leaps and bounds in its framework which can go from public to private, for its algorithms that can even run on IoTs, which helps in remote connectivity, for its distributed network that has information of every transaction made by a peer from time to time, the update of information is a strength in this technology, and the main reason we are foreseeing this technology as a future lead in networking domain is its security which is in the hands of every peer and sophisticated rule to chain a block. The minimal use of third parties is a default advantage of blockchain. The Blockchain in tech improvements are going to make this world of technology a secure place for the functioning of operations on the surplus amount of information that is being produced for every second.

This technology will be used as a communicating platform between devices in an authorized manner which is seen as a better service provided in every sector.

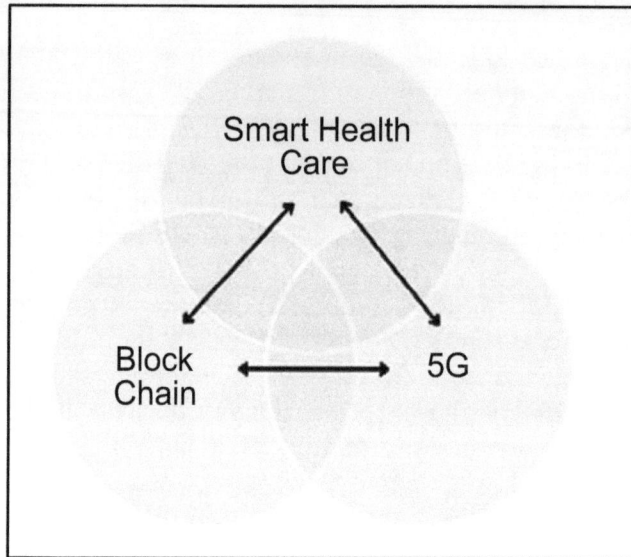

Fig. (1). Venn representation of blockchain.

Block-Chain

The innovations considering both upcoming and existing programs in Blockchain can project the existing technology in a peculiar and efficient manner. Blockchain is a shared and connected immutable ledger that records transactions and updates them in a frequent manner. The sailors of the technology are eagerly exploring this island intermixing with all other techs such as IoTs, IoMTs, IIoTs, finance, banking, healthcare, and AI which gives us untapped possibilities in an efficient manner. Although Block-chain was introduced as a financial manager to save the clients having trust issues and third-party organizations, this is seen as the front-liner of every network communication because of its security. Before getting into the very enlarged and flexible domain, let us understand its basics as shown in Fig. (**2**).

Block is a basic unit of a Block-chain, which contains three essential things for its connectivity 1) data is the information we want to transact. 2) Hash-key of the block is the fingerprint through which we can access the information of the block (32 bit,64 bit,128 bit,256 bit). 3) Previous Hash is the hash-key value of the previous block that's implemented in the chain as [1]. There are some crucial points to be noted about a block such as, the information in a block is immutable or intangible and it can hold the info of user choice, the chains in the block are introduced so that no other malicious block enters in between the chain and corrupt the data, the ownership can be changed in hand to hand along with the transactions, which are entirely time-specific and permission-restricted.

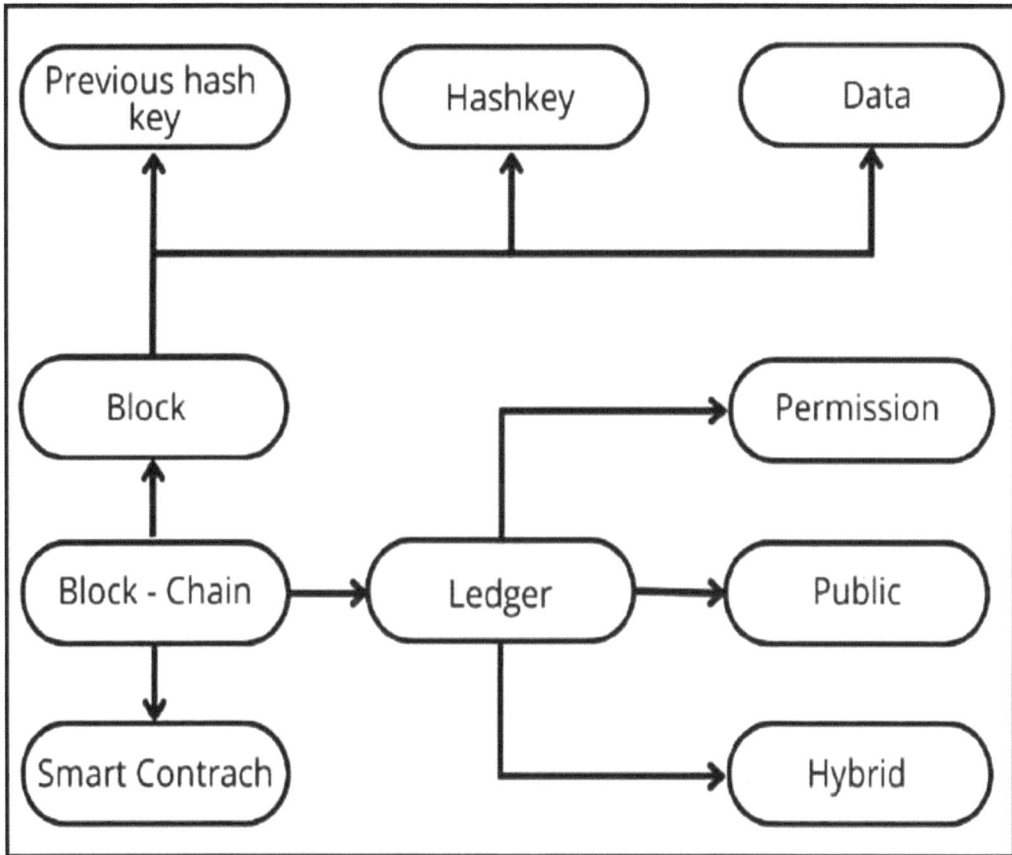

Fig. (2). Mechanism of blockchain.

Block-chain is majorly dependent on three key elements:

Distributed Ledger

This forms a distributed network replacing the centralized network before, this technology majorly helps in maintaining uniform information all over the network which is controlled through a distributed number of nodes.

There are mainly three types of distributed ledger technologies:

- **Permissioned:** Permission should be taken from an entity by using our access key given by them Eg: Banks, Healthcare.
- **Permissionless:** This is a public blockchain, anyone may create a block. Eg: Crypto, Ethereum.
- **Hybrid Permissioned:** This is a combination of the above technologies, which provides the benefits of each of them. There are different DLTs available such as block-chain, hash-graph, holochain, tempo, and radix.

SMART CONTRACTS

Smart contracts are basically programs that are recorded on a blockchain and run when certain criteria are satisfied. Smart contracts work in a way where they can terminate their conditions after transactions or completion of the time period, it is independent of additional or intermediary delay. Smart contracts operate in such a way that they can automate the process of workflow when "if/then...when" phrases are used in block-chain and written in the code [2]. A network of computers will execute these operations whenever predefined conditions have been verified to have been satisfied. When the process is completed and the blockchain is updated, to determine whether the outcome circumstances are met, this can only be verified by the parties with permission granted, and the following action is executed.

Block-Chain and its Applications

Block-chain in Drug Supply Management

The medication chain of supply has frequently been capable of providing solutions for human issues in the fields of healthcare and medicine. The issue is a crucial component with regard to the knowledge of security, information sharing techniques, and buying that medicine. Sadly, illicit drugs that are purchased from markets put sick individuals at risk of dying since they are false, making the importance of drug safety vital. The pharmaceutical and medical sector has changed to digitization and the internet in recent decades. So the medical community came to the solution of blockchain. BCT helped IT and ICT become more powerful. In reality, BCT with smart contracts can improve the industry's or drug's tracking and level of confidence. BCT is a safe way to exchange information [3].

Blockchain in Agriculture

Numerous areas of agricultural systems are improved by blockchain applications, particularly the Internet of Things (IoT) and supply chain-based systems. Food security, quality, safety monitoring and control, food tracking to reduce waste, effective data processing and data analysis, support for small-scale farmers, food security, and efficient contract exchanges and transactions are some of these applications [4]. These apps can be created quickly and easily by utilizing the blockchain platforms already in use. Multiprocessor and cryptography processes can be combined to give flexibility and fulfill high user expectations given the various deployment situations for these applications [5].

Blockchain in Banking

Transactions in a banking system must go through a bank, hence the approach does not include the usage of miners or tokens. No one else, even though they are a part of the system, has the authority to add a block, only the bank has this power. The banking system will check, validate, and record the transaction if person A needs to transfer money to person B before adding a new transaction to the block. The system uses the SHA-256 algorithm. Secure Hashing Function or SHA [6, 7] communicates the numerical value of the fixed piece length, and the numerical count of 256 does the same.

Block-chain in IoTs

Utilizing block-chains as the network control mechanism for the IoT will boost its flexibility because they are distributed databases irrespective of the semantics. The diversity of IoT devices and protocols currently limits their capacity to communicate with one another. In order to meet the rising needs and desires of IoT users, a blockchain-based IoT framework promises to adapt to various settings and use cases. Blockchains have been demonstrated to work over heterogeneous hardware platforms. The proliferation of constantly accessible smart gadgets that gather data and offer automated functionality is what is meant by the IoT. High availability is required by network control protocols for the Internet of Things (IoT) [8], the systems using centralized servers may not always be the case. Through distributed consensus protocols, block-chains are robust fault-tolerant tracking systems that are able to identify errors. Fig. (**3**) explains these applications.

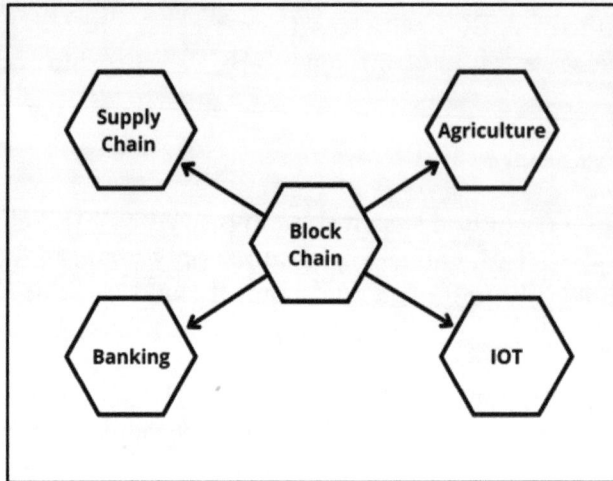

Fig. (3). Application of blockchain in different sectors.

Smart Health-Care System

The challenges that are being faced in the healthcare sector have significantly increased as a result of an increase in diseases and patients. The health sector and services are highly essential in the present situation even for the people of very remote areas but the system is lagging behind with its old technology. These rising rates of illnesses and patients present particular difficulties. These difficulties may include increased healthcare expenses, a lack of medical staff, pressure on drug pricing, *etc.*, resulting in inadequate operation of the healthcare system. The below problems are studies based on [9, 10] as shown in Fig. (4).

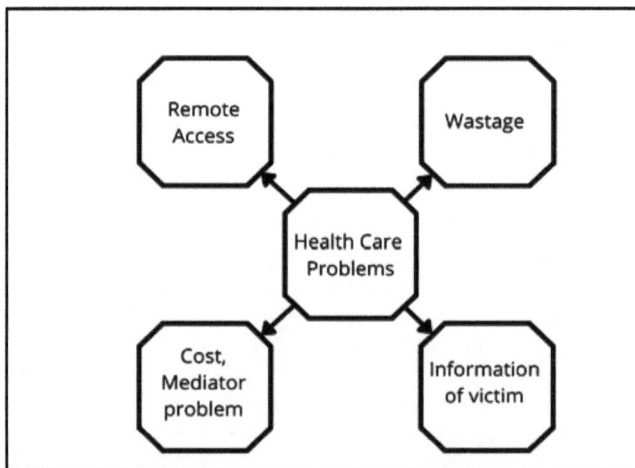

Fig. (4). 5G and its technology.

The problems with the present health care system:

Wastage

This is a problem that has to be addressed globally because of its high effect on nature and human beings. One of the leading waste dumpers in all domains is the health sector. From small pins to syringes, from bandages to tablets, from dresses to old technologies, there are some kilotons of waste every month. During this COVID time, additional wastes are added like masks, sanitizers, *etc.* Although the recycling is being done at a very fast rate, that is not enough.

Information of the Victim

This is the most underrated and overlooked problem in the medical industry. The patient or victim enters the hospital with one problem and leaves the hospital with another. There are a lot of cases where the patients die due to abnormalities because of improper treatment. This is because of the lack of records of the patients and they are not updated from time to time. When people have accidents and unexpected medical concerns, these situations become significantly worse.

Cost and Mediator Problems

Most of the medicine buyers do not know why the costs are that heavy in hospitals, it is because of the mediators and contractors in the supply chain. Most of us also do not know why and where the costs are applicable. This is why we spend more in hospitals and lose trust in them.

Remote Access to Medicines and Treatment

The present healthcare does not provide health access to people in remote areas and they have to travel a lot for each visit. Sometimes the patients or victims when meeting with accidents in some tourist places, should wait for the rescue for more hours, this is because of a lack of connectivity between the patient and the nearest hospital.

So some of the problems of our present health care system are being discussed above. We can clear all these issues by making the healthcare system smart enough, like the entry of the data in the system, and maintaining them from time to time record. The smart-health care system is not decentralized and anyone can breach the network and can change the data of the patient, which causes fatal medical errors. So these kinds of problems are addressed using blockchain, which will be discussed in the entire chapter.

Introduction to 5G-NR

5G is a fifth Generation Communication Technology that is going to revolutionize the world and change the parameters of communication and networking such as latency, security, quality of services, throughput, inter-device communication, variable data rates and many more to a new set of improved versions as shown in Fig. (**5**). The major service providers of 5G are Reliance Jio, Ericsson, AT&T, Verizon, *etc.*, 5G principle use cases are divided into three groups, which are Enhanced Mobile BroadBand (eMBB) *i.e.*, various ranges of Bandwidths this technology enables, Ultra Reliable Low Latency Communication (URLLC) which explains the latency of 5G that is reduced even if it includes more devices than in 4G and finally about machine type communication.

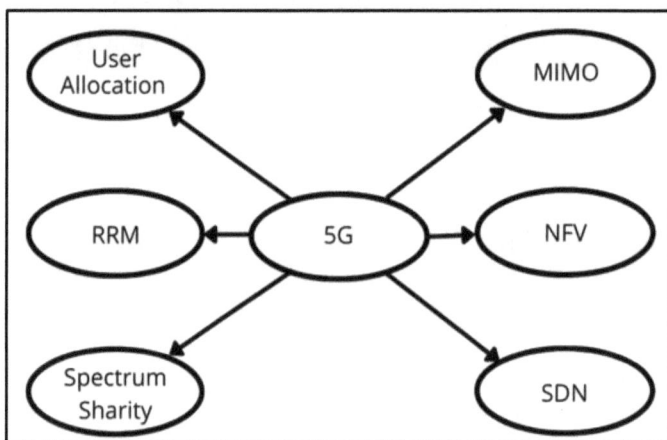

Fig. (5). Issues related to healthcare sector.

The evolution of 5G from 4G has occurred with the Radio Resource Management, which briefs about the reliable methods, algorithms for observing energy transmissions, modulation schemes, and line codes that should be enabled for effective communication considering the power management, congestion control techniques, and allocation of resources. These RRM techniques are used in spectral management, resource allocation to the users, providing the computational interfaces and efficiency, better throughput, and also better security with the distributed networks. This RRM is mainly based on the MIMO technique using multiple 2D array antennas for longer and variable bandwidths [11].

Mainly SDN and NFV play a major role in forming the 5G distributed networks. These are used in the standardization of control interfaces and also provide good policies for maintaining the network and avoiding the degradation of performance and help in controlling the network. 5G in the blockchain is very efficient because of its crowdsourcing techniques data connected to multiple registered operators and customers this distributed network of 5G heads blockchain per automatic implementation of payments and the links to be tracked transparent manner and also helps in better sharing of smart contracts which are already established connections another advantage is the spectrum sharing where it is known as trustless environment and blockchain can become a crucial role in offering the tracking of ownership and usage in a secure manner a decentralized application can be used for all these performances finally there are other advantages like network slicing and international roaming can be integrated for organizations to be created for their own network [12].

Introduction to MEC

The massive increase in mobile network demand prevents network expenditures from matching profits, referring to the network resources used by modern enterprises and customers are placing numerous restrictions, challenging mobile communications sectors like personalized services, enhanced functionality, and user experience. However, the mobile operator is still facing the effects of declining ARPU(Average revenue per user), which influences their being mostly confined to the worst bit pipe providers struggling with optimizing operations, resource management, and developing creative services that generate income while maintaining the standard of service quality. This flawed business strategy could be the reason for the unprecedented growth and innovation.

By putting IT and cloud computational power close to mobile users in the RAN, MEC (MEC) is a viable option for the aforementioned problems. It offers a cutting-edge network architecture in which a cellular network for application delivery is combined with the availability of cloud computing and a setting for IT services. Its benefits come from the distinctive qualities that set it unlike a standard cloud platform in the following ways: Severe closeness to the user (seldom do more than 1 or 2 network hops), which allows for significantly reduced core network congestion and decreased final delay. Application developers are exposed to the radio network's capabilities, including their connection with operator network networks and access to network data. In order for the mobile user to quickly access the data, the mobile network must provide accelerated content delivery. In the MEC-like platform, several measures are taken to address it, enhancing data transmission by introducing TCP performance enhancement proxy (PEP) is a middle box between both the base station and the core network (CN). With the radio network supplying a seeming real-time data on the radio channel capacity into the TCP packet's option field, the mobile terminal obtains data transmission commands from the TCP server located outside the phone service.

The mobile network must unload IP traffic in accordance with the pre-set rules in order to serve applications close to the mobile user. The 3GPP has put out a number of methods for the traffic offloading process known as local breakout. A UE linked through a HeNB can access objects' IP capability within the IP network using the Local IP Access (LIPA) solution without having to go *via* the mobile operator's network. The UE requests are based on an APN and establish a PDN connection. The network chooses the L-GW connected to the HeNB in order to allow a direct path in the user plane between the L-GW and the HeNB. Different redundancy-based protection strategies, including 1: 1 and 1: N, can support ultra-dependability. Each strategy places certain restrictions on the MECs but also gives the network a certain capability, in order to support trustworthiness. As per the studies, the 1:1 system can have higher reliability levels, but at a substantially higher price. The 1: N plan, on the other hand, offers a more economical deployment. Taking use of earlier discoveries, This is an additional possibility of a 1: N: K protection plan to increase the 1's capacity 1: N scheme [13].

DIFFERENT BLOCK-CHAIN DESIGNS FOR SMART HEALTH-CARE

Platforms for Secure Tracing of Drugs

Using Blockchain in B2B

After registering the site qualification in the block-chain distributed network, the B2B (business-to-business) e-commerce platform, a blockchain node, regularly realizes the transaction data on the chain. The upstream businesses, or drug distribution businesses, positioned on the B2B platform must fill in their business details, including business information like license no of drug business, generalized system of preference certificate, *etc.*, such as medication-related documents like the registration and licensing of the different dispersed drug types as shown in Fig. (**6**). Drug retailers who buy through B2B e-commerce platforms must also submit the credentials of their business. Retailer organizations that purchase medications using the platform are supposed to digitally register their orders, including the name, batch number, details, volume, and person in charge of the drug procurement. After the creation of the upstream and downstream purchase orders, the platform itself creates smart contracts, ensures information security through asymmetric encryption, and eliminates contract or trust breaches by recording the data of both parties to the transaction information into the blockchain. To achieve tracking and traceability in real time, logistics businesses, which are accountable for transferring information between drug distribution businesses and drug retail businesses, must enter data of logistics like the transportation environment and drug journey nodes, into the logistical procedure promptly [14].

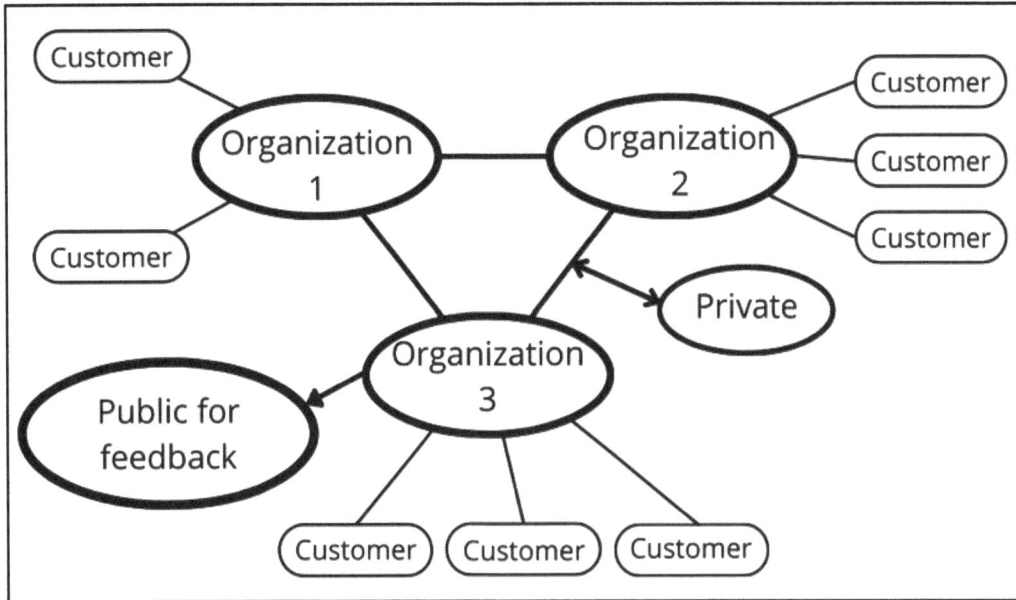

Fig. (6). B2B – blockchain.

Using Blockchain in B2C

The business-to-business and business-to-customer (B2C) e-commerce platforms are similar and dissimilar. The eligibility of the platform and the business requirements of the upstream B firms, including business license and business permission, are identical in terms of registration, storage, and information uplink. When a user purchases drugs using the platform, details about the transaction and the drugs, such as the identity of the purchaser, the warehouse and shelf number where the drugs are stored, the logistics unit number, *etc.*, are recorded. These details will be uploaded concurrently once the order is formed. The logistics firm in charge of drug transportation will submit the records, and those requiring special transportation will also be documented and uploaded with environmental conditions by specialist logistics firms. In order to fill up the health condition and prescription information after passing the verification, the B2C platform must first gather the identification details of the drug user who orders prescriptions. Visit the website for validating name information on national ID cards to accomplish this. The following information will be updated synchronously on the chain using asymmetric encryption once the prescription drug purchaser has been audited for adds tracking and subjected to scrutiny by pertinent departments. Blockchain technology is used in the aforementioned connection to ensure simple pharmaceutical traceability in this mode and help monitor the flow of prescription drugs [15].

Privacy Sharing Technology with the Block-chain

Framework

The privacy sharing architecture includes four main categories of businesses, or subjects: Source Data (terminal source of data), Industry Partnership, Data Internet Provider, and Data Buyer. Privacy of user contracts, user permission agreements, industry partnership regulatory norms contracts, and privacy data transaction contracts all manage interactions between entities. The general procedure is as follows: after signing the user privacy data contract, the user submits to the service provider their private information. The user can also establish precise usage guidelines for uploading private data by using the authorization contract. Additionally, industry alliance regulators impose limitations on data exchanges between data service providers through industry-standard contracts.

Storage with Respect to the Cipher Text

The personal data collecting program will gather user privacy data from terminals such as cars, smart devices, cameras, and mobile terminals according to the user's privacy policy and any applicable industry or national privacy compliance rules. According to the particular data type, software collection can determine whether to encrypt or desensitize the data throughout the gathering procedure. Collected data can be sent to the data service provider to process it. To address the issue of privacy and data leak during transmission, relay nodes are implemented. To lessen the danger of privacy leaking in the next stage, the private data may be combined or mixed up by the relay node. After obtaining the private data, the data service provider has two options for storing it: either on their systems or with a private data storage provider. Some essential technologies are required to address the issue of privacy data loss and leakage during storage [16]. The encryption text of several storage nodes is stored using RAID (Redundant Arrays of Independent Disks), and secure multi-party computation technology is utilized to do certain privacy data converging computation for machine learning to increase the value of the data.

The Methodology for Implementation

Privacy data sharing through hybrid chain architecture, which can improve data exchange privacy, data fusion supervision, and concurrently take into account the efficiency of blockchain systems. It is challenging to strike a good balance between consensus security, cost, and efficiency built on a single chain topology for managing privacy sharing. As a consequence, a mixed chain architecture was created, consisting of the alliance chain and the public chain. The alliance chain is in charge of recording transaction accountability and exchanging private information, and its nodes are made up of data service providers and independent assessment [17]. Off-chain storage system can strike a solid privacy/scalability balance. All data cannot be stored directly on the chain due to the current sluggish processing speed of the blockchain. In order to tackle computer congestion during rush hour, the off-chain storage strategy is implemented to link the already-existing blockchain and shift the complex and personal computing to the off-chain network. In order to protect shared secret data, blockchain employs decentralized hash tables as the controller for this storage computer network [19].

Block-chain Methodology for Supply Chain in Healthcare

The supply chain management system is made up of a number of stages and different supply chain application industries. Supply chains are more dependable thanks to blockchain technology. The POM (Product Ownership Management) system that is presented encourages counterfeiters to replicate real products because they are not able to establish ownership of those products under that system. Then they developed a system that permits the transfer and ownership proof of RFID (Radio Frequency Identification) tags at each step of the supply chain, allowing users to manage inventories and identify products automatically and uniquely. Small businesses nowadays frequently face financial challenges that are a fewin comparison to those faced by major corporations with robust financial resources. Small businesses in the brand management industry will inescapably need to cut expenses [18].

A decentralized approach using blockchain technology ensures that users do not rely on the merchants to determine whether things are genuine. As a result, the producer can implement this approach to supply actual products without having to oversee the independently owned storefronts. It can significantly lower costs and ensure product quality. The technique can successfully reduce the bar for authentic goods and give businesses with minimal financial resources. Additionally, it is a simpler strategy to give customers the assurance that they will not buy fraudulent goods. There isn't any code redundancy or simplicity, though. The Maker Chain Decentralized Application (DApp) reveals how individual makers may self-organize around one another.

The latest decentralized supply chain utilizes blockchain and Near Field Communication (NFC) technology to detect attacks on counterfeit items. In addition to having technology for product tracking and traceability, the block supply chain can detect efforts at tag duplication, moderation, and reapplication. They created a brand-new, safe, and scalable protocol just for this chain. For huge networks, it is incredibly trustworthy and effective. It will be a good option for massive blockchains that demand complete centralization. An adversary may decide to ignore the fact that the number is unchanging.

The details about the product that the producer made are contained in the QR code. The exchange of goods and services between the seller and the buyer is safe and unbreakable. A distinct QR code will be included in every blockchain transaction that cannot be used by the producer on another product. A blockchain-based management (BCBM) system can be used to record product information and its unique code as database blocks. The transaction block is inaccessible to unauthorized parties. Any participant who requests product information must provide the manufacturer with their public key. The QR code is encrypted by the manufacturer and sent back to the participant. By using their private key, the legitimate participant will decrypt the QR code.

Interoperability of Health-Care

Networks that support population health initiatives can access massive amounts of patient data using the blockchain thanks to data interoperability undertakings. Patients benefit from having a transparent leader because they have access to their own medical information as well. The following can be done to achieve healthcare data interoperability and observance of legal and regulatory requirements. A blockchain for health care may also be connected to the most recent scientific findings to help patients choose therapies. Consequently, patients will be better informed and speak to their doctors about the best course of treatment based on data and not authorized techniques. Through the creation of secure and reliable health record information, the connection of transactional data, and the provision of patient access and consent, blockchain technology is also helpful for the interoperability of healthcare information. By utilizing advanced APIs(Application Programming Interfaces), it is possible to deliver safe EHR interoperability and data storage, fulfilling the core promise of blockchain technology. The cost and duty of data reconciliation will be eliminated with the steady and uniform sharing of the blockchain network to approved providers. Otherwise, blockchain will alter how revenue cycles, drug supply, clinical trials, and fraud prevention are managed.

Different methods and technologies involved in Block-chain for interoperability are:

Polkadot-Polkadot

Is a well-known blockchain that takes interoperability to a new level. The aim of Polkadot, one of Ethereum's creators, is to advance the exchange of smart contract data between blockchains. Transactions can be broadly dispersed because

Polkadot has numerous blockchain network channels. Absolute secrecy is upheld in all business dealings. The aim of Polkadot is to create a safe connection between public networks, private chains, and open oracles. The proponents of blockchain interoperability solutions aspire to make it possible for various blockchain systems to communicate efficiently with each other *via* a Polkadot relay network [20].

Aion-Online

Another well-known block-chain interoperability project, Aion Online aims to solve unresolved scalability and interoperability problems in block-chain networks.

BlockNet

Developers are presently constructing a decentralized exchange to enhance interchange communication. Blockchain also uses interoperability block-chain techniques, which could change how we now see blockchain. Blocknet intends to decentralize all four of these components in order to establish the first decentralized exchange. The cross-chain network's infrastructure is also optimized by the project's backers.

Cosmos Block-chain

Is the most underappreciated block-chain interoperability effort. The block-chain project is currently at the center of a number of initiatives. A software development kit that tackles the scalability and interoperability issues in blockchain applications has since been released by the creators. A hub is the core block-chain of the cosmos architectural block-chain, which is made up of several distinct zones called block-chains. Both a high-performance and reliable consensus engine are similar to the Practical Byzantine Fault Tolerance (PBFT) model. Through the inter-block-chain communication protocol, the cosmos Hub links block-chain projects to enhance interoperability [21].

Blockchain and IOMT for Remote Access to Smart-HealthCare

The integration of medical data (such as administrative records, clinical testing data, ailment archives, and health information), insurance entitlements, health surveys, *etc.* in a coordinated, cost-effective manner that supports self-government and saves time and effort is made possible by the conjunction of blockchain and IoMT technologies. This is in addition to enabling remote patient monitoring. In order to diagnose and effectively treat diseases like cancer and cancer of the internal organs, as well as to make good use of sensory devices like ingestible medications, heart rate monitors, and other medical equipment, block chain technology is used.

A summary of some of the healthcare benefits of blockchain technology IoMT as observed from is as follows [20]:

Collaboration

It helps with effective management and synchronization using distributed ledger technology, and the field's innovation offers individualized medical treatment in times of need in Acute Care Facilities.

Data Provenance and Integrity.

It enables the processing and storage of information for the growing number of users and devices in healthcare institutions.

Data Protection

This takes care to protect sensitive papers and information and forbids unauthorized people from accessing data.

Monitoring

This permits access to medical data, documents transactions in a clear and transparent manner, and reduces time, labor, and expense.

Process Simplification

This reduces the amount of work required to protect sensitive information, which enhances the system as a whole.

Enables the Realization of Smart Hospitals

Even finding someone or anything saves time on administrative work and makes catalog management duties more effective, which lowers the cost for healthcare service providers.

Provides Individualized Medical Treatment

Including in intensive care units and emergencies (ICU).

The several ways to integrate blockchain technology into IoMT have been divided into categories. The various categories include ethereum, contributions based on

Hyperledger, general blockchain ideas without technical specifications, modified cryptographic techniques, and modified consensus protocols.

Ethereum-Based Contributions

An Ethereum-based architecture aims to remotely monitor diabetic patients, with data access being managed *via* smart contracts. The consensus process is implemented through smart contracts in Malamas' own Ethereum-based architecture. The suggested architecture makes use of three Smart Contract (SC) characteristics to enable registration [Registration Smart Contract (RSC)], permission [Actor Handling Smart Contract (AHSC)], and logging requests [Log Management Smart Contract (LMSC)]. It also directs the players' cries to the technical or medical resources. All of the health-related information is stored in an Inter Planetary File (IPFS). To defend against harmful behavior, a Proof of Medical Stack (PoMS) is suggested. Khatoon suggested a private Ethereum-based blockchain to manage medical data. Data access between all parties is provided through Ethereum smart contracts. Along with imaginative depictions of medical records, the smart contract also contains metadata regarding record ownership and data quality. To protect data integrity, the cryptographic hash is maintained on the blockchain, whereas the health information is stored off-chain (on an external server). A cloud-based IoMT framework was created by Nguyen *et al.*, using an ethereum-based blockchain network, to transfer and share data securely among clients in the healthcare industry.

Hyperledger-based Contributions

Attia *et al.* suggested an IoT-blockchain-based architecture to enable remote health monitoring. Blockchain is divided into two categories: consultation blockchain and blockchain for medical devices. In order to visualize the patient's health data, a user interface is created. Chain codes in fabric are used to substantiate and authenticate the transactions.

Modified Consensus Protocol

is a consortium blockchain-based architecture proposed by Uddin. To define the BlockChain features, the authors created patient agent software (PA), which was implemented on the Edge computing network. Health data handling may be done using smart contracts in numerous ways, including filtering out clinically worthless data, setting alarms for certain circumstances, moving data to the cloud as necessary, categorizing data, and more.

Modified Crypto Technique

For encryption, Natarajan *et al.* used a cutting-edge encryption method along with a hashing technique. The method supports the IoMT's real-time responsibility while covering all medical objects with very little temporal complexity. To get around the POW consensus prototype, Dwivedi *et al.* customized and privatized the blockchain-based framework. IoMT devices generate a large amount of data, which is gathered as encoded data in blocks and stored in the Cloud. On the blockchain, block hashes are recorded. To ensure anonymity, Uddin *et al.* suggested a ring signature, which is utilized as a common public key-based digital signature [21].

The General Block-chain Concept without Technical Specifications

Blockchain tamper-proof feature was used by Gupta *et al.* to securely store and share the IoMT data. The IoMT data is saved in blocks in an off-chain database, while the patient data is accumulated as blocks in the blockchain. The privacy and security of the blockchain are guaranteed by smart contracts. A blockchain-based structure for the MedChain consortium was suggested by Shen *et al.* to share data streams produced by medical sensors effectively. The blockchain network and the P2P network are two distinct, decentralized sub-networks of the MedChain network. The BFT-SMaRt is the consensus protocol used by MedChain. The BIoMT, an efficient, lightweight blockchain-based framework, was proposed by Seliem *et al.*

5G IN BLOCK-CHAIN USING ALGORITHMS AND ITS APPLICATIONS

In 2019, the fifth-generation (5G) wireless technology was implemented, and in 2020 it was used commercially. In comparison to current 4G LTE networks, next-generation 5G wireless communications are anticipated to offer extremely high data rates (typically in the Gbps range), extremely low latency, a significant increase in base station capacity, and a significantly better customer perception of service quality (QoS). 5G is anticipated to link people, objects, data, apps, healthcare systems, cities, and nations in surroundings with smart networked communication. New applications, such as augmented and virtual reality, smart cities, 3D movies, cloud computing, and work, as well as blockchain-secured medical infrastructure and databases are also used. It sparks a revolution that transforms everything, no matter how tiny. With these requirements in mind, the 5G network was created. Mobile apps and connected devices require wireless network access that is dependable, secure, and able to safeguard users' privacy. The security of the 5G infrastructure, like the electrical and energy infrastructures on which we presently rely, would, nevertheless, have an impact on how well the society as a whole runs. As we rely on 5G more, security threats increase.

Although 5G network employs cryptographic techniques to defend itself against the great majority of hostile adversarial attacks, it is created with a more sophisticated and robust security architecture than the 4G LTE network.

5g Systematic Approach

The applied systematic technique is built on the model-checking notion. Model checking is an automated verification method for examining complicated systems that state transition systems can model. It exhaustively enumerates (explicitly or implicitly) all the states that the system is capable of reaching and all the actions that traverse between them in order to demonstrate that the reactive system exhibits the specific behavioral feature across the given system (the model). The model-checking approach has the following notable benefits over simulation, testing, and other formal verification techniques [22].

It is completely automated, requiring neither user supervision nor expertise in mathematical fields like logic and theorem proving. The model-checking approach reliably generates a counterexample showing behavior that falsifies the required attribute when a design does not fulfill it. This flawed trace offers priceless information into the real reason for the failure and vital hints for fixing the problem.

These two important potential advantages, along with the invention of symbolic model checking, which permits thorough implicit enumeration of a massive number of states, have fundamentally transformed the study of formal verification. From being a completely academic topic, it has evolved into a useful, doable methodology that may be used as an extra useful tool for design validation inside various industry development processes [23]. The approach to model verification is consequently of great interest to the academic community. It has also been used in other technological sectors to check the system's security and dependability. (1) Modeling: In order to represent synchronous communication with finite-state machines, we must first choose the network components to be employed in compliance with the appropriate technique of the examined network protocols from 3GPP standards. The protocol model is then abstracted in accordance with the functionality of the evaluated network protocols as specified in the 3GPP specification. We choose an adversary model in the second step. In order to create the threat model instrument, we finally changed the protocol's ability to estimate the presence of an opponent. (2) Extract the desired property in step two. To highlight the essential method of the examined network protocols and identify what characteristics we need, we must unite 3GPP standards. The desired characteristics are then extracted from the 3GPP specifications.

Discover Concerning Attacks: First, we inserted the desired attribute retrieved from the threat model instrument into the model checker. Second, we checked if any model executions violated the property using the model checker. If a model execution were to be violated, this would serve as a counter example. If not, we determine that the model satisfies the property [24]. We must identify this counter-example using a cryptographic protocol verifier called CPVerify as it may be false due to the abstraction.

Orthogonal frequency division multiplexing is an efficient modulation technique used in contemporary wireless communication systems, notably 5G. (OFDM). OFDM combines the benefits of Quadrature Amplitude Modulation (QAM) with Frequency Division Multiplexing (FDM) to produce a high-data-rate communication system (QAM). QAM encompasses a variety of modulation techniques, including BPSK (Binary Phase Shift Keying), QPSK (Quadrature Phase Shift Keying), 16QAM (16-state QAM), 64QAM (64-state QAM), and others. Simply said, FDM is the notion that several communication channels may coexist if a portion of the frequency spectrum is set aside for each channel. FM broadcast radio is a typical illustration of this: the total (US) frequency allotment ranges from 87.8 MHz to 108 MHz and is split into channels that are 0.2 MHz wide (Fig. **1**). To reduce interference from neighboring channels, FDM frequency allocations frequently feature guard bands between the channels and must not overlap.

For any blockchain to be mentored properly, it requires security and more transactions per second with less computational power so the 5G algorithm is proposed on this basis it has highly varying data speeds that can be observed from the above information and it has more security rules than the previous one with more encrypted and decryption techniques incorporated in it. The main use of 5G in the blockchain is its power to have remote control access. This can be done using Iot where the devices communicate with each other and share the information in the most secure way using blockchain algorithms and contracts. One more advantage of 5G properties above is that it is not centralized like previous generations, it is distributed all over and uses the MEC for faster computational speeds, which helps the Health service respond quickly with the additional techniques of Blockchain for security even in the remote areas.

Blockchain as a Service Platform for Local 5G Operators

In the L5GO ecosystem, the BaaS architecture functions as an overlay entity. The BaaS architecture's proposed block-chain-based services will be delivered through an overlay block-chain. These two distinct strategies can be used to implement blockchain: both a consortium blockchain and a public blockchain. The current blockchain platforms (like Ethereum) can be used to implement the public blockchain. The BaaS architecture's suggested services should be put into practice. However, this is pricey because of the expense of running as the value of digital money rises. Consequently, a consortium blockchain for each stakeholder is mirrored in the BaaS architecture (MNOs, L5GOs, VNF providers, *etc.*) node.

Each participant's needs may be catered for in the blockchain deployment paradigm. The deployment configuration, for peer nodes and mining nodes, for instance, may be customized based on the requirements. Both MNOs and L5GOs may function as miner nodes for peer transactions and mining. Due to their high capacity, VNF and cloud service providers can be used as miner nodes. Fog computing nodes, which might not have as much processing capacity as IoT nodes, can be used to deploy the matching blockchain nodes. The blockchain nodes of the IoT tenants can only be utilized in this case as full nodes that carry out transactions that have been committed to the network.

Each stakeholder's blockchain node is capable of performing the customized services required by the system. An IoT ant colony's blockchain node, for instance, may manage IoT data that is shared with other services *via* smart contracts. The two primary benefits of integrating blockchain nodes to deliver the services are the ability to manage relatively greater quantities of transactions compared to cloud-oriented designs and eliminating delay by the local blockchain node. The use of a cloud service, which entails a data transit leg over the Internet, causes a bottleneck when the system gets a greater amount of transactions. The blockchain node delivers perimeter security by enabling service deployment closer to the stakeholder.

The main functions of the BaaS architecture are:

The Subscription Management Function (SMF)

This process is to register stakeholders and service applications with the system. The service management role is suggested to help with registration. Taking the subsequent actions enables the system to register stakeholder information as well as numerous application-related resources on the blockchain [25].

Marketplace Function (MF)

This concept intends to create a marketplace where sellers can advertise their products and customers can conveniently make purchases. Smart contracts may be used to combine several segment algorithms and bidding processes with this function in order to choose the best product that is currently accessible. Smart contracts may be used to automate the selection process.

Reputation Management Function (RMF)

The effectiveness of the services provided to the various stakeholders is assessed using our approach. The relative importance of the stakeholders will be evaluated using this historical performance data, and the payment rates for their services will be determined. Thus, we provide a cutting-edge reputation management tool for assessing the goods and services provided by the network suppliers. The main function of this reputation management tool is reputation scoring. It may be used primarily for roaming and offloading services for each network provider. It is also used to evaluate how well the market's merchants are doing.

Selection Function (SF)

In addition to selecting the best L5GOs to complete an offload or roaming assignment, the system must carry out a number of other selection duties. As a result, we propose a selection function that would allow the system to automatically pick the optimal network provider for a mobile user performing a roaming and offloading event [26]. When an offload occurs, MNOs must select the best subscriber or subscribers to unload. The selection mechanism can therefore choose the ideal subscriber for offloading.

Fraud Prevention Function (FPF)

The purpose of the FPF is to prevent fraud from having an impact on roaming and offloading occurrences. It focuses specifically on preventing visiting users from using too many resources.

Data Management Function (DMF)

L5GO networks typically include a variety of Internet of Things (IoT) devices. Other users can access the collected IoT. We suggest a data management function that is primarily focused on the administration of data access and storage, two crucial facets of IoT data management.

The complete summary of the above services provided by BaaS is unique and appreciable with a combination of every architecture to make a smart healthcare system in 5G using blockchain a more secure, selective, easily managed, and better way of sharing the right information based on the reputation of a node in the blockchain. This service is a mixture of both private and public consortium algorithms, which makes it more unique for good case studies in blockchain about patients in a more transparent manner. The architectures that are used such as the Local 5G operating network which will be remotely accessed help blockchain with Iot to control and look after medical services even in remote areas [27, 28].

There are a lot of contracts that are seen in this blockchain such as user registration contract, network registration contract, offload decision contract, network selection contract, network reputation management contract usage limit contract, cost calculation contract, seller registration contract, product registration contract, search product contract, product purchase contract, and seller reputation management contract, which completely take the smart healthcare system to a new place [29, 30].

Federation Algorithm for Block-chain using 5G

Explanation of the System Architecture

Imagine an extensive dispersed-edge server and node network that is wireless. The edge nodes are users of cars, cellphones, and IOT devices who have specialised processing and communication capacities. Base Stations or Macro Base Stations (BSs or MBSs) and MEC servers make up the edge servers. The suggested system is composed of components such as regional training, a blockchain for parameter validation, and worldwide aggregation [28]. The edge servers are responsible for upholding the blockchain and performing the global aggregation procedure. Edge users do the local training using their data, where Di belongs to D. At the beginning of federated learning, edge servers post the worldwide model M0's parameters w(0) to the blockchain as a starting point for learning. The edge nodes execute local training and obtain the global parameters w(t) from the blockchain in order to ascertain the changes Delta of w(t) for Mt on the data sources at iteration. To lessen the likelihood of updated parameter leak, privacy protection techniques like differential privacy and asymmetric encryption can be used prior to sending the updates. The Delta of w (t) changes are then sent to the blockchain after that. The blockchain-based delegated proof of stake (DPoS) consensus technique assesses the precision of changed parameters. The BSs distribute transactions to selected users for verification and gather the findings of statistical verification. The verifiers evaluate the model's quality based on their collected local data, once the parameters have been determined through the consensus method.

Asynchronous Federated Learning

Due to the diverse processing capabilities and various communication statuses of heterogeneous devices, conventional federated learning may create considerable delays in maintaining synchronization between participating nodes. We propose an asynchronous federated learning approach that simultaneously takes into account computing and communication resources. We also consider updates' characteristics, which depend on the local data and computing environment. the successful asynchronous educational strategy. The efficacy of federated learning may be increased by focusing on the following parts in particular, which can be based on control strategies to resolve the specified goal function.

- **Node Selection:** The efficiency of federated education is primarily influenced by the nodes that will be participating because of the varied abilities and supplies of various clients. Additionally, it is important to confirm the developed models of every node. In order to maximize learning performance depending on computing capabilities, communication status, and training data quality, the algorithms for selection control can be improved.
- **Global Aggregation:** Selecting the first global model for learning in line with the application conditions is a critical challenge for the network edge. Further research is required in order to create the algorithm for aggregation, choose the length of time for every iteration, and determine several iterations because the aggregating process consumes a large number of resources [29].
- **Local Training Trade-off:** The participating nodes seek to achieve more training accuracy at a lower cost, hence there is an exchange because blockchain-based solutions are time-consuming mechanisms for upgrading changes. The control methods for choosing the number of regional training iterations prior to submitting the updates can further boost the profit of the regional training process.

BLOCKCHAIN EMPOWERED FEDERATED LEARNING

The blockchain parameter links disparate participants including servers that aggregate data and nodes. The edge users might be consumers or producers of resources. The blockchain gathers all resources and requests for sharing, together with details on their amount and time frame. The inquiries are sent to MBS and then broadcast to potential suppliers, who subsequently reply to MBS with information on the tools they have at their disposal, such as computers and interactive resources. The MBSs then distribute the requests to the providers and carry out the learning process utilizing federated learning and blockchain. Three phases make up the method: local training, global aggregate, and parameter blockchain.

Deep Reinforcement Learning

For node selection, we take advantage of shared educational control. Client nodes are encouraged to participate in the federated education if they have higher computer resources, connection capabilities, and data quality, according to node selection. The learning process may exclude nodes with insufficient resources in order to enhance overall performance. The edge server, which also acts as the learning system's aggregator, is where the DRL server is housed. DRL develops the optimum model by engaging with the environment; it does not require any training data or model assumptions. Three essential elements of DRL are the state, the action, and the reward mechanism. The three crucial elements of our DRL are the primary network, target network, and repeat memory. A pair of Deep Neural Networks (DNNs) that are a component of the primary network, which also has a structure with the intended network, are the actor DNNs with crucial DNN. The target network's values of interest are used to train the main critic DNN. By linking model parameters to actions, the actor DNN investigates the rules. The critic's DNN calculates the policies' efficacy, which leads behavior in the direction of the policies' gradient.

The experience reminisces approach reduces the volatility brought on by incorrect predictions. The experienced tuples, which are kept in the replay memory, contain the present state, selected action, reward, and following state. The DNNs are trained using the knowledge tuples in a randomly sampled method, which may disentangle the behavioral connections between various training episodes.

It is a combination of Deep Learning with Neural Networks that is being developed in 5G on the basis of a Federation Algorithm, which learns every node information in the chain and awards it based on its state. From the chains added, it learns the additional data that can be added to the nodes, which are encouraged by the organizations; these algorithms will tune its parameters based on every node and any malicious info of health data centers into the chain can be easily identified by the algorithm and can be removed.

CONCLUSION

The chapter introduces the scope of blockchain in various fields from agriculture to different services, it also goes into the history of blockchains and why they were first established in the introduction part. This chapter explains the blockchain from the very start where we can see the nodes and their properties, along with the smart contracts, which are used to make blockchain function for transactions. This paper also explains the different types of blockchains like public, permissioned, and hybrid models that are widely used.

Different applications of blockchain have been explained in this chapter such as its use in drug supply, agriculture, banking, and IoT in which the blockchain is used for connectivity in a distributed and secure way with an additional advantage of transparency and remote access. In recent times, most of the Fintech companies started using blockchain such as J.P Morgan, Bajaj FinServices, Asian Banks *etc.* There are a lot of companies in the world with the combinations of IoT-agriculture and Blockchain such as Xage, Helium, NetObjex, and Atonomi who provide security to IoT devices with the help of Blockchain and improve their efficiency.

Further introduction of smart healthcare systems is also provided which explains how important is the patient's information in critical times. It also explains how the information should be shared and updated for a better suggestion among the community using blockchain and also explains the mediator problems that can be resolved using the blockchain. Then we introduce 5G techniques such as NOMA, QAM, and OFDM, which help in creating variable internet speed in the blockchain on top of it, the transactions per second are increased in a drastic way when compared to the existing 4G technology. MEC Technology plays a major role in 5G, which is explained along with 5G services such as Baas and Federated algorithms, which use Local 5G operators and are further connected in the cloud at a very global level. It continues to explain the various services that are provided by the blockchain that can be implemented such as supply chain management of drugs, Blockchain in privacy sharing using Hyper Ledgers and also explains the various interoperability techniques of the blockchain. Finally, this chapter is concluded with an explanation of 5G services that are provided for increasing transactions, providing better security, and creating better virtualization functions that can remove the computing power using MEC services and can also increase market capacity to all places providing equal services of health care.

REFERENCES

[1]   S. Lin, Y. Kong, and S. Nie, "Overview of block chain cross chain technology", *13th International Conference on Measuring Technology and Mechatronics Automation (ICMTMA)*, pp. 357-360, 2021.
[http://dx.doi.org/10.1109/ICMTMA52658.2021.00083]

[2]   L. Wei, and W. Lanjia, "The Nature of block chain intelligent contract", *15th International Conference on Computer Science & Education (ICCSE)*, pp. 805-810, 2020.
[http://dx.doi.org/10.1109/ICCSE49874.2020.9201768]

[3]   I. Al Barazanchi, "Blockchain: The next direction of digital payment in drug purchase", *International Congress on Human-Computer Interaction, Optimization and Robotic Applications (HORA)*, pp. 1-7, 2022.
[http://dx.doi.org/10.1109/HORA55278.2022.9799993]

[4]   Shivendra, K. Chiranjeevi, M. K. Tripathi and D. D. Maktedar, "Block chain Technology in Agriculture Product Supply Chain", *International Conference on Artificial Intelligence and Smart Systems (ICAIS)*, pp. 1325-1329, 2021.
[http://dx.doi.org/10.1109/ICAIS50930.2021.9395886]

[5]   B. Hegde, B. Ravishankar, and M. Appaiah, "Agricultural supply chain management using blockchain technology", *International Conference on Mainstreaming Block Chain Implementation (ICOMBI)*, pp. 1-4, 2020.
[http://dx.doi.org/10.23919/ICOMBI48604.2020.9203259]

[6]   N.S. Akhilesh, M.N. Aniruddha, and K.S. Sowmya, "Implementation of blockchain for secure bank transactions", *International Conference on Mainstreaming Block Chain Implementation (ICOMBI)*, pp. 1-10, 2020.
[http://dx.doi.org/10.23919/ICOMBI48604.2020.9203095]

[7]   N.P. Pravin, K.P. Anil, S.M. Sunil, M.S. Kundlik, and P.A. Suhas, "Block chain technology for protecting the banking transaction without using tokens", *Second International Conference on Inventive Research in Computing Applications (ICIRCA)*, pp. 801-807, 2020.
[http://dx.doi.org/10.1109/ICIRCA48905.2020.9183333]

[8]   J.H. Jeon, K-H. Kim, and J-H. Kim, "Block chain based data security enhanced IoT server platform", *International Conference on Information Networking (ICOIN)*, pp. 941-944, 2018.
[http://dx.doi.org/10.1109/ICOIN.2018.8343262]

[9]   U.T. Khan, and M.F. Zia, "Smart city technologies, key components, and its aspects", *International Conference on Innovative Computing (ICIC)*, pp. 1-10, 2021.
[http://dx.doi.org/10.1109/ICIC53490.2021.9692989]

[10]   V. Tripathi, and F. Shakeel, "Monitoring health care system using internet of things - an immaculate pairing", *International Conference on Next Generation Computing and Information Systems (ICNGCIS)*, pp. 153-158, 2017.
[http://dx.doi.org/10.1109/ICNGCIS.2017.26]

[11]   J. Lee, M. Han, M. Rim, and C.G. Kang, "5G K-SimSys for Open/Modular/Flexible system-level simulation: Overview and its application to evaluation of 5g massive MIMO", *IEEE Access,* vol. 9, pp. 94017-94032, 2021.
[http://dx.doi.org/10.1109/ACCESS.2021.3093460]

[12]   S. Manap, K. Dimyati, M.N. Hindia, M.S. Abu Talip, and R. Tafazolli, "Survey of radio resource management in 5g heterogeneous networks", *IEEE Access,* vol. 8, pp. 131202-131223, 2020.
[http://dx.doi.org/10.1109/ACCESS.2020.3002252]

[13]   Z. Jia, D. Li, W. Zhang, and L. Pang, "5G MEC gateway system design and application in industrial communication", *2nd World Symposium on Artificial Intelligence (WSAI)*, pp. 5-10, 2020.
[http://dx.doi.org/10.1109/WSAI49636.2020.9143280]

[14]   S. Narang, M. Byali, P. Dayama, V. Pandit, and Y. Narahari, "Design of trusted B2B market platforms using permissioned blockchains and game theory", *International Conference on Blockchain and Cryptocurrency (ICBC)*, pp. 385-393, 2019.
[http://dx.doi.org/10.1109/BLOC.2019.8751472]

[15]   L. Wanganoo, B. Prasad Panda, R. Tripathi, and V. Kumar Shukla, "Harnessing smart integration: Blockchain-enabled B2C reverse supply Chain", *International Conference on Computational Intelligence and Knowledge Economy (ICCIKE)*, pp. 261-266, 2021.
[http://dx.doi.org/10.1109/ICCIKE51210.2021.9410677]

[16]   X. Deng, T. Lv, and L. Song, "Novel efficient block chain and rule-based intelligent privacy share system in future network", *Conference on Computer Communications Workshops (INFOCOM WKSHPS)*, pp. 1-8, 2022.
[http://dx.doi.org/10.1109/INFOCOMWKSHPS54753.2022.9798148]

[17]   J. Noh, and H. Kwon, "A Study on smart city security policy based on blockchain in 5G Age", *International Conference on Platform Technology and Service (PlatCon)*, pp. 1-4, 2019.
[http://dx.doi.org/10.1109/PlatCon.2019.8669406]

[18]   I. Al Barazanchi, H.R. Abdulshaheed, Z.A. Jaaz, H.M. Gheni, Y. Niu, H. Almutairi, E. Daghighi, S.A. Shawkat, and S.R. Ahmed, "Blockchain: the next direction of digital payment in drug purchase", *International Congress on Human-Computer Interaction, Optimization and Robotic Applications (HORA)*, pp. 1-7, 2022.
[http://dx.doi.org/10.1109/HORA55278.2022.9799993]

[19]   M.C. Jayaprasanna, V.A. Soundharya, M. Suhana, and S. Sujatha, "A block chain based management system for detecting counterfeit product in supply chain", *Third International Conference on Intelligent Communication Technologies and Virtual Mobile Networks (ICICV)*, IEEE., pp. 253-257, 2021.
[http://dx.doi.org/10.1109/ICICV50876.2021.9388568]

[20]   J. Indumathi, A. Shankar, M.R. Ghalib, J. Gitanjali, Q. Hua, Z. Wen, and X. Qi, "Block chain based internet of medical things for uninterrupted, ubiquitous, user-friendly, unflappable, unblemished, unlimited health care services (bc iomt u 6 hcs)", *IEEE Access,* vol. 8, pp. 216856-216872, 2020.
[http://dx.doi.org/10.1109/ACCESS.2020.3040240]

[21] A.K. Das, B. Bera, and D. Giri, "Ai and blockchain-based cloud-assisted secure vaccine distribution and tracking in iomt-enabled covid-19 environment", *Internet of Things Magazine,* vol. 4, no. 2, pp. 26-32, 2021.
[http://dx.doi.org/10.1109/IOTM.0001.2100016]

[22] P.R. Kapula, J.G. Jeslin, G. Hosamani, P. Vats, C.J. Shelke, and S.K. Shukla, "The block chain technology to protect data access using intelligent contracts mechanism security framework for 5g networks", *2nd International Conference on Advance Computing and Innovative Technologies in Engineering (ICACITE),* pp. 202-206, 2022.
[http://dx.doi.org/10.1109/ICACITE53722.2022.9823471]

[23] Z. Chen, S. Chen, H. Xu, and B. Hu, "A security authentication scheme of 5g ultra-dense network based on block chain", *IEEE Access,* vol. 6, pp. 55372-55379, 2018.
[http://dx.doi.org/10.1109/ACCESS.2018.2871642]

[24] H. Ko, J. Lee, H. Choi, and S. Pack, "Hierarchical Identifier (HID)-based 5G Architecture with Backup Slice", *21st Asia-Pacific Network Operations and Management Symposium (APNOMS),* pp. 291-293, 2020.
[http://dx.doi.org/10.23919/APNOMS50412.2020.9236966]

[25] N. Weerasinghe, T. Hewa, M. Liyanage, S.S. Kanhere, and M. Ylianttila, "A novel blockchain-as-a-service (BaaS) platform for local 5g operators", *IEEE Open J. Commun. Soc.,* vol. 2, pp. 575-601, 2021.
[http://dx.doi.org/10.1109/OJCOMS.2021.3066284]

[26] V. K. Rathi, "A blockchain-enabled multi domain edge computing orchestrator", *IEEE Internet of Things Magazine,* vol. 3, no. 2, pp. 30-36, 2020.
[http://dx.doi.org/10.1109/IOTM.0001.1900089]

[27] M. Saravanan, S. Behera, and V. Iyer, "Smart contracts in mobile telecom networks", *23RD Annual International Conference in Advanced Computing and Communications (ADCOM),* pp. 29-33, 2017.
[http://dx.doi.org/10.1109/ADCOM.2017.00011]

[28] Y. Li, C. Chen, N. Liu, H. Huang, Z. Zheng, and Q. Yan, "A blockchain-based decentralized federated learning framework with committee consensus", *IEEE Netw.,* vol. 35, no. 1, pp. 234-241, 2021.
[http://dx.doi.org/10.1109/MNET.011.2000263]

[29] Y. Lu, X. Huang, K. Zhang, S. Maharjan, and Y. Zhang, "Blockchain and federated learning for 5g beyond", *IEEE Netw.,* vol. 35, no. 1, pp. 219-225, 2021.
[http://dx.doi.org/10.1109/MNET.011.1900598]

[30] D. Pradhan, "5G-green wireless network for communication with efficient utilization of power and cognitiveness", *InInternational Conference on Mobile Computing and Sustainable Informatics: ICMCS,* vol. 2021, pp. 325-335, 2020.

CHAPTER 6

Edge Computing for Analysis in Health Care Industry using 5G Technology

B. Sahana[1,*], Dhanush Prabhakar[1], C.S. Meghana[1] and B. Sadhana[2]

[1] *Department of Electronics and Communication, R. V. College of Engineering Bangalore-560059, India*

[2] *Department of Electronics and Communication, Canara College of Engineering, Mangalore, India*

Abstract: In today's world, ailments have increased due to increased stress and an unhealthy way of living among other reasons. This demands proper and effective monitoring of an individual's health for early prevention. Among the various ailments, heart-related issues have become a significant concern. The increased risk of heart-related problems can be tackled by the use of technology, which provides a route for effective monitoring, therefore various ways pertaining to technologies have been explored. Extensive research has been conducted in the fields of smart textiles and sensors, with Textile Electrocardiogram being one of the major developments. Electro-cardiography (ECG) is a popular technique for monitoring the heart rate and other parameters in order to alert the individual of any risk if present. However, real-time monitoring is crucial for reliable and effective analysis. This analysis can further be converted into reports for proper diagnosis by certified medical professionals or doctors. Adequate and efficient analysis of this data requires enormous resources and computing power, which implies that mobile phones are not suited for the same. This leads to the necessity for customized hardware to achieve this task. In view of this, an architecture has been developed to interface the sensors wirelessly using 5G protocols for faster and secure communication to the custom Hardware *i.e.* edge device to generate reports on demand. In this chapter, we will discuss the recent advances in various technologies that can be used at the communication, encryption and edge computing levels, the challenges, and potential solutions.

Keywords: Edge computing, Edge device, Smart healthcare, Textile electrocardiogram, 5G communication.

INTRODUCTION

A distributed computing paradigm is known as edge computing [1]. Many issues, such as the longevity of a battery's charge limitations, response time requirements,

---

* **Corresponding author B. Sahana:** Department of Electronics and Communication, R. V. College of Engineering Bangalore-560059, India; E-mail: sahanab@rvce.edu.in

Devasis Pradhan, Mangesh M. Ghonge, Nitin S. Goje, Alessandro Bruno and Rajeswari (Eds.)

bandwidth, cost, and data security and privacy, could be resolved *via* edge computing. Some of the challenges in edge computing include security and privacy, resource management, and programming models. Edge computing allows for the low-cost implementation of improved performance systems, which can be applied to newer fields. Edge computing represents a dynamic and evolving ecosystem, characterized by fragmentation. Moreover, both standards and business models are presently evolving and reaching a phase of maturity.

Edge computing applications are vastly supported by the introduction of 5G communication systems [2], which provide high bandwidth, low latency, and massive connectivity. Technology will play a bigger part in healthcare in future, and 5G-based smart healthcare networks will be essential to enabling cutting-edge medical applications. These networks do, however, also present a of security difficulties.

Integrating 5G connectivity into the IoT systems introduces the risk of connectivity and reliability issues as many devices can be connected to the same network at some time. The impact of 5G on healthcare extends beyond just faster internet [3]. 5G networks present numerous benefits for healthcare applications, positioning themselves as a promising technology in the field. With their high data transfer rates and minimal latency, 5G networks facilitate real-time communication and swift data transmission, essential for efficient healthcare services. These networks facilitate the seamless transmission of high-resolution medical imaging data, enabling remote diagnosis and consultation [4, 5].

The emergence of 5G technology enhances the existence of a highly interconnected Internet of Things, primarily through Massive Machine Type Communication (mMTC). This demands robust network capacity to handle numerous connected devices within each cell, thereby generating the need for effective communication with control centers [6]. 5G network ecosystem facilitates seamless connectivity between medical sensors, actuators, and the cloud, offering exceptional high-speed and extensive bandwidth support [7].

Edge computing, by relocating computation and storage closer to the data origin, facilitates quicker analysis and decision-making, a critical aspect in healthcare applications [8]. The development of numerous technologies that interface sensors wirelessly using 5G protocols has been made possible by notable advancements in wireless sensors for wearable electronics in recent years [9]. The integration of wearable sensor arrays for in situ perspiration analysis allows for large-scale application. This integration allows for complex signal processing and wireless transmission, connecting the technological disparity between signal transduction, conditioning, processing, and wireless communication in wearable biosensors [10].

The significant enhancement of application performance and the processing of extensive real-time data are enabled by the interconnected technologies of 5G and edge computing. Speeds are increased up to ten times by 5G compared to 4G, while latency is reduced by mobile edge computing, as it brings computing functionalities nearer to the final user within the network. The use of edge computing and 5G in healthcare applications comes with challenges [10, 11].

This chapter will cover the growth of edge computing, its growth since the arrival of 5G communication technology, and its impact on the health sector. Each of these concepts and their impacts on the society will be discussed.

OBJECTIVES OF EDGE COMPUTING WITH 5G

Leveraging edge computing within 5G technology profoundly amplifies the efficacy of wearable devices like ECG devices. This integration empowers real-time analysis, swift responses, and streamlined data transmission, fundamentally advancing the oversight and control of cardiac health for individuals.

Fig. (1) illustrates the five objectives of edge computing in 5G for healthcare application considering an example of Textile Electrocardiogram.

Fig. (1). The objectives of Edge computing with 5G.

For example, a person wearing a connected ECG device experiences a sudden change in heart rhythm. The edge computing capabilities of the device allow for immediate analysis and recognition of the anomaly. Instead of transmitting all the raw ECG data, only the pertinent information regarding the abnormal heart rhythm and associated details are sent to the healthcare provider or hospital systems for further analysis.

Improving Data Management

Wearable ECG devices continuously monitor a person's heart activity, generating a substantial amount of data. Edge computing enables on-device processing to identify critical patterns (like arrhythmias or abnormal heart rates) without sending the entirety of raw data to the cloud. This reduces bandwidth usage and allows for faster response times in detecting and addressing cardiac irregularities.

Improving Quality of Service (QoS)

Real-time monitoring and processing at the edge help maintain a high level of service quality. For instance, in the case of an ECG device detecting a potentially life-threatening arrhythmia, immediate alerts can be sent to healthcare providers or emergency services for rapid intervention, thus improving patient outcomes.

Predicting Network Demand

Analysing ECG data at the edge helps predict when more network resources might be required. For instance, during peak times when multiple ECG devices are simultaneously sending data, edge computing can anticipate these spikes in data transmission, ensuring the network is prepared to handle the increased demand without compromising the data's critical nature.

Managing Location Awareness

Wearable ECG devices integrated with edge computing can provide real-time information about a patient's heart status along with their precise location. In emergencies, this location-awareness capability can aid in directing emergency medical services to the patient swiftly, enhancing response times and potentially saving lives.

Improving Resource Management

Edge computing optimizes resource utilization by processing and storing critical data on the device itself. Only crucial information, such as detected anomalies or irregular heart rhythms, is transmitted to the healthcare facility or the cloud. This minimizes the load on network resources and central servers while ensuring that essential data is efficiently managed.

This utilization of edge computing in 5G technology significantly enhances the effectiveness of wearable ECG devices by enabling real-time analysis, prompt action, and efficient data transmission, ultimately improving the monitoring and management of cardiac health for individuals [11, 12].

Edge Computing Analysis

Edge computing has numerous applications in present-day IoT industry. With the proliferation of the internet, it is becoming essential for devices to compute locally to avoid delays caused by transmission or communication. Edge computing is the solution to decrease latency and network congestion by extending cloud computing to the edge of the network. Edge computing enables real-time analysis and control of streaming data generated by IoT and mobile applications [12, 13].

Some of the Applications of Edge Computing include

Edge computing is applied in various scenarios, such as smart systems power monitoring, and power metering systems. It can be used in the power monitoring systems to keep track of power usage and control of the power usage in a particular area. It can also be used to make smart decisions on power or energy usage based on previous data and information [13, 14].

It is essential in building 5G networks and supporting the infrastructure for scalable deployment of edge computing nodes. Edge computing allows the low-performance devices on the edge to control 5g connectivity at particular edge clouds, which work together to control the connectivity on a larger scale [14, 15].

Deep learning has been applied in edge computing applications and a thorough examination of the present research in this domain has been carried out. Edge computing allows us to use AI to control edge systems. Integration of deep learning allows for high-quality decision systems at the edge devices, hence increasing the reliability of the system [15, 16].

5G COMMUNICATION ANALYSIS

The rise of 5G connectivity has led to increased adaption of edge computing technologies in many fields. With faster and more reliable communications systems, edge computing has spearheaded itself as the forerunner of embedded system design.

Mobile edge computing (MEC), a pivotal technology within the burgeoning fifth-generation (5G) network, optimizes mobile resources. It achieves this by hosting compute-intensive applications, processing substantial data before transmitting to the cloud, furnishing cloud-computing abilities within the radio access network (RAN) situated close to mobile users, and delivering context-aware services utilizing RAN information [16].

5G Connections are seen to be Effective in the Following Fields

Mobile Health: 5G's low latency and high bandwidth allow for real-time monitoring of patients. Health data from mobile devices allows for remote consultations and interventions.

Augmented Reality (AR) Surgery: 5G's high bandwidth enables the transmission of high-quality video streams, allowing surgeons to use AR headsets to overlay medical images onto the patient's body, enhancing surgical precision.

Haptic Feedback: 5G's low latency enables real-time haptic feedback, allowing surgeons to feel the resistance of tissues and organs during surgery, improving surgical dexterity and control.

Remote Patient Care: 5G's high bandwidth enables real-time video conferencing and remote consultations, allowing healthcare providers to monitor and care for patients remotely.

Connected Ambulances: 5G enables the transmission of medical data from ambulances to hospitals in real-time, allowing for faster diagnoses and treatment upon arrival [17].

Edge Computing and 5G Analysis in Health Care

The edge computing architecture can be used in the healthcare industry with the help of 5G communication to enhance the performance and reliability of the entire system. The present-day healthcare system consists of a lot of machinery which are interconnected and require a lot of computing prowess. The introduction of edge computing-supported 5G communication promises an exponential leap in this industry.

Some of the places where this technology can be implemented are as follows:

Real-time Patient Monitoring: Edge computing can be used for collecting and analyzing patient data in real-time, such as vital signs, blood pressure, and glucose levels. This data can be used to detect potential health problems early on and provide timely interventions. By leveraging edge computing capabilities, data from patient monitoring devices can undergo processing and analysis in real time at the edge of the network, reducing latency and enabling an immediate response [18]. This is particularly crucial in emergency situations where timely intervention is critical. One example of a proposed framework is the FairHealth system, which utilizes 5G edge computing to provide Healthcare focused on long-term proportional fairness within the system. Internet of Medical Things (IoMT)aims to optimize the allocation of healthcare resources and ensure fair access to healthcare services by leveraging the capabilities of 5G edge computing. Furthermore, edge computing and 5G together can enable the seamless integration of various healthcare devices and systems. Wearable sensors, medical devices, and other IoT devices can transmit data to edge computing nodes, where it can be processed and analyzed in real time. This diminishes the necessity for data transmission to centralized servers, improving efficiency and addressing privacy and security concerns [20].

Telemedicine: Edge computing can enable telemedicine by providing the low latency and high bandwidth required for real-time video conferencing and remote consultations. With the application of 5G networks, the effect of video transmission in telemedicine can be improved, leading to better quality and more reliable remote consultations. Additionally, the increased bandwidth of 5G networks allows for the transmission of large medical imaging files, enabling remote interpretation and consultation [21]. The integration of edge computing and 5G in telemedicine also enables the use of Internet of Medical Things (IoMT) devices [22, 23]. These devices, such as wearable sensors and remote monitoring devices, generate a vast amount of data that can be processed and analyzed at the edge, reducing the need for data transmission to centralized servers . This not only improves the efficiency of data processing but also addresses privacy and security concerns by keeping sensitive data within the local network [22, 23]. The integration of edge computing and 5G technology holds great potential for advancing telemedicine. The combination of real-time communication, low latency, and efficient data processing enables high-quality remote consultations, remote monitoring, and remote interpretation of medical imaging [23, 24]. However, challenges related to security, infrastructure deployment, and privacy need to be addressed to fully realize the benefits of telemedicine using edge computing and 5G.

Medical Imaging: Edge computing can accelerate the processing of medical images, such as X-rays, CT scans, and MRIs, allowing for faster diagnoses and treatment decisions. Medical imaging plays a crucial role in diagnosing and monitoring various medical conditions. The integration of edge computing and 5G technology has the potential to enhance medical imaging processes, enabling faster and more efficient image acquisition, transmission, and analysis. Edge computing can significantly improve medical imaging by processing image data closer to the source, reducing latency, and enabling real-time analysis. This is particularly important in time-sensitive scenarios, such as emergency situations or surgical procedures, where immediate access to high-quality images is critical for making informed decisions. By leveraging edge computing, medical imaging devices can offload computational tasks to local edge servers, reducing the burden on centralized cloud infrastructure and ensuring faster processing times [23, 24]. The integration of 5G technology further enhances medical imaging capabilities by providing high-speed and low-latency communication. With 5G, medical imaging devices can transmit large image files quickly and reliably, enabling real-time collaboration between healthcare professionals and facilitating remote consultations. The increased bandwidth and capacity of 5G networks also support the seamless transmission of high-resolution images, improving the accuracy and quality of diagnoses [24]. Furthermore, the combination of edge computing and 5G enables the implementation of advanced image analysis techniques, such as artificial intelligence (AI) algorithms, directly at the edge. AI-powered edge devices can perform real-time image processing and analysis, allowing for automated detection of abnormalities and assisting radiologists in making accurate diagnoses [25]. This not only improves the efficiency of medical imaging workflows but also enhances the overall quality of patient care. The integration of edge computing and 5G technology holds great promise for advancing medical imaging capabilities. The combination of real-time image processing, high-speed communication, and AI-powered analysis at the edge can significantly improve the efficiency, accuracy, and accessibility of medical imaging. However, addressing security and privacy concerns and ensuring seamless integration are crucial for the successful implementation of these technologies in medical imaging workflows.

Surgical Robotics: Edge computing can improve the performance of surgical robots by reducing latency and improving responsiveness, leading to more precise and less invasive surgeries. The use of 5G networks in surgical robotics enables high-speed and reliable communication between the surgeon and the robotic system. This is crucial for remote surgeries, where the surgeon may be located in a different location from the patient. The low latency provided by 5G ensures that the surgeon's commands are transmitted to the robotic system in real-time, allowing for precise and immediate control [26]. Additionally, the high bandwidth of 5G networks facilitates the transmission of high-definition video and imaging data, providing the surgeon with a clear and detailed view of the surgical site [26, 27]. The combination of edge computing and 5G in surgical robotics opens up new possibilities for remote surgeries and telemedicine. Surgeons can remotely control robotic systems with haptic feedback, allowing for precise and delicate procedures to be performed from a distance [27]. This is especially beneficial in situations where specialized expertise is not readily available in certain locations. The use of 5G telesurgery has been demonstrated in urology, showing the potential for remote surgeries to be conducted safely and effectively [28]. The integration of surgical robotics, edge computing, and 5G technology holds great promise for advancing the field of surgery. The combination of real-time communication, low-latency control, and enhanced data processing capabilities enables remote surgeries and telemedicine, bringing specialized expertise to underserved areas. However, challenges related to network reliability, security, and infrastructure deployment need to be addressed to fully realize the potential of surgical robotics using edge computing and 5G.

Wearable Devices: Edge computing serves to process data from wearable devices, such as [29] smartwatches and fitness trackers, providing valuable insights into a patient's health and activity levels. Edge computing plays a crucial role in wearable devices by enabling local processing and reducing the need for data transmission to centralized servers. This results in lower latency, improved privacy, and reduced dependence on cloud infrastructure [30]. 5G technology complements edge computing in wearable devices by providing high-speed and low-latency communication capabilities. It enables seamless connectivity between wearable devices and other healthcare systems, facilitating real-time data transmission and remote monitoring. Wearable devices can leverage edge computing to perform real-time data analysis, such as monitoring vital signs, detecting anomalies, and providing timely feedback to users [30, 31]. Additionally, wearable devices can support personalized fitness and wellness applications, providing users with real-time guidance and motivation to achieve their health goals. Wearable devices leveraging edge computing and 5G technology have the potential to transform healthcare and personal well-being. These devices can enable real-time monitoring, personalized interventions, and remote collaboration between users and healthcare professionals. However, challenges related to security, power consumption, and interoperability need to be addressed to fully realize the benefits of wearable devices in healthcare [31].

EDGE COMPUTING ARCHITECTURES IN 5G TECHNOLOGY

SDN-Based Edge Computing

SDN-Based Edge computing is structured as a three-layer framework, consisting of the infrastructure layer, an edge computing layer, and a core computing layer. It provides centralized control and management of network resources, enhances mobility and QoS management, supports network slicing and routing, and enables the implementation of security enhancements.

It facilitates network slicing, enabling the development of virtual networks with tailored features and attributes and resources [31, 32]. SDN-based architectures shown in Fig. (**2**). can support efficient handover signaling and integrated solutions for LTE Evolved Packet Core (EPC) and 5G networks [32]. SDN can also be used to create intelligent core networks for the tactile internet and upcoming smart systems [33].

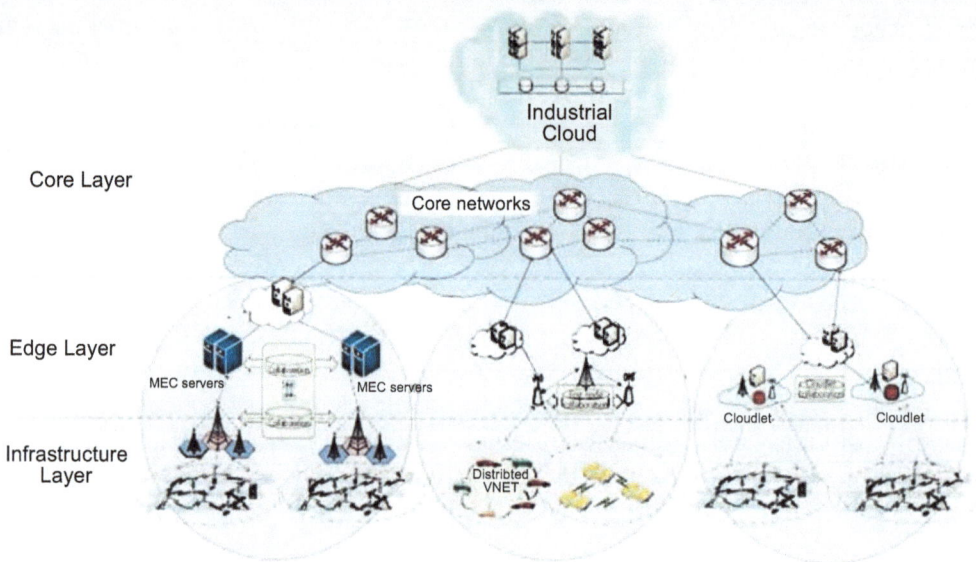

Fig. (2). The three layer architecture of SDN-based edge computing [34].

The edge computing layer consists of a network of SDN-based edge servers that are responsible for intelligent data processing, storage, and seamless communication. SDN-enabled Edge computing strives to provide a range of quality of service (QoS) benefits, including low latency, quick response times, improved efficiency, increased throughput, and proximity-based services. To achieve these goals, an efficient resource allocation system is integrated within the SDN controller, ensuring optimal network configuration and load balancing. The controller possesses comprehensive knowledge of each Edge server's storage, processing, and communication capabilities, enabling it to manage incoming and outgoing traffic from IoT devices effectively. Based on server loads and predetermined rules, the SDN controller orchestrates the cooperation between edge servers, facilitating the efficient utilization of computational and storage resources through their coordinated efforts [34].

Layers of SDN Edge Computing Architecture are as Follows

Infrastructure layer: The healthcare IoT infrastructure layer encompasses low-powered embedded sensors and diverse IoT devices used for patient monitoring. These devices, varying in operating systems, CPU, memory, and transmission powers, operate in hospitals or on the patient's body.

Edge computing layer: The Edge computing layer consists of numerous edge servers catering to the high volume and diverse functionality of IoT and wearable devices. These servers manage data exchange, storage, processing, and job migration.

The core computing layer: The core computing layer encompasses the core networks and cloud services responsible for hosting diverse applications and managing the end-to-end IoT architecture. Equipped with robust data protection measures, these layers require implementation of authentication, authorization, and cryptographic mechanisms.

Application-centric Design of 5G and Edge Computing Applications

The application-centric design of 5G and edge computing for healthcare involves a shift in network architecture towards provisioning networks tailored according to the needs of individual applications [35].

The application-centric design of 5G and edge computing is supported by technologies like network slicing, software-defined networking, and network function virtualization. Network slicing allows the formation of customized virtual networks designed for specific applications., allowing for the efficient allocation of network resources. Software-based networking and network function virtualization provide flexibility and programmability in managing and deploying network functions, further enhancing the application-centric approach [36].

A critical abstraction layer referred to as the "app slice" offers a unified approach considering both the compute resources essential for applications and the intricate network resources. This tier efficiently employs container management systems to supervise edge computing resources and utilizes the 5G network stacks for managing network resources shown in Fig. (**3**).

"App slice," enables the combined definition of the computing and network prerequisites for the entire application and its individual functions. Runtime system actualizes the specified declarative semantics described in the app slice specifications, facilitating unified management of Leveraging two adaptive algorithms to manage edge computing and network resources in various edge-computing environments and 5G networks. The specification at the application level entails parameters such as latency, bandwidth, device count, and reliability. Simultaneously, the function-level specification focuses on network attributes, including latency, throughput, packet error rate, and duration. Computing requirements for the functions are detailed within the compute specification, encompassing minimum and maximum CPU cores and memory resources.

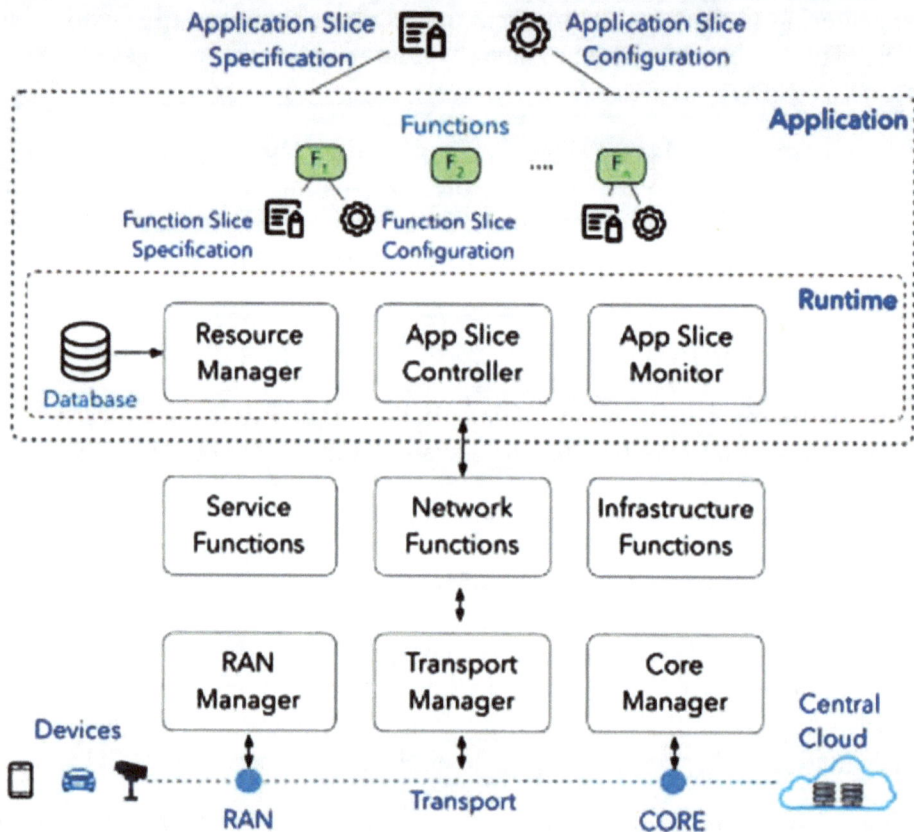

Fig. (3). 5G and edge computing architecture using application-centric approach [37].

PERFORMANCE WAS OBSERVED TO INCREASE USING APP SLICE [37]

Offloading Computation

The offloading of computation tasks from IoT devices to nearby edge servers or 5G edge is a key technique to improve the performance and efficiency of IoT applications. This offloading process involves transferring computationally intensive tasks from IoT devices to edge servers with higher computational capabilities, reducing the burden on devices and improving their energy efficiency.

The offloading sequence encompasses multiple phases. Initially, the IoT device identifies tasks suitable for offloading based on specific criteria like task complexity or edge server availability. Subsequently, the device establishes communication with the nearby edge server or 5G edge to commence offloading. The task is then transferred, computed by the edge server, and the output is returned to the IoT device.

Determining which tasks to offload and their destination depends on factors such as edge server capabilities, network conditions, and IoT device energy limitations. Implementing optimization algorithms and machine learning aids in making informed offloading decisions and dynamically adjusting to evolving conditions [38, 39].

The Mobile Edge Computing Architecture

The Mobile Edge Computing (MEC) framework is a Framework that extends cloud computing services to the edge, specifically to mobile base stations. It consists of components such as MEC hosts and the MEC platform, along with enablers like Network Function Virtualization and network slicing. Combining MEC with 5G networks increases the capability of edge computing and enables advanced applications and services [39]. NFV allows for the virtualization of network functions, enabling flexible and scalable deployment of services on the MEC hosts. Network slicing, on the other hand, enables the creation of virtualized network instances tailored to specific applications or services, ensuring efficient resource utilization and isolation [40].

ADVANTAGES OF USING EDGE COMPUTING WITH 5G COMMUNICATION

Criteria for Real-time Processing

Edge computing in tandem with 5G results in reduced latency, which is vital for instantaneous healthcare applications [40, 41]. This capability enables swifter data processing and decision-making, especially critical in urgent scenarios such as emergency medical care [41].

Data Transmission Capacity

The extensive bandwidth offered by 5G enables the swift and efficient transmission of substantial data, facilitating the utilization of high-resolution medical imaging and remote patient monitoring [41, 42].

Security and Privacy

Storing and processing data at the network's edge keeps sensitive medical information closer to its origin, thereby reducing the vulnerability to data breaches [42, 43]. Implementing edge computing can facilitate the adoption of zero-trust architectures, significantly bolstering the security of healthcare systems [44].

Device Longevity Reliability

Leveraging edge computing alongside 5G enables the transfer of computational tasks from medical devices to the edge, lessening the workload on the devices. This action extends their battery life, facilitating continuous monitoring without the need for frequent recharging or battery replacements in the devices [44, 45].

Independent from Cloud Reliability

The fusion of edge computing and 5G facilitates deploying healthcare services nearer to the point of care, diminishing the dependence on centralized cloud infrastructure [45]. This advancement can result in enhanced scalability and responsiveness of healthcare systems, alongside a decrease in network congestion [46].

Low Cost

Edge computing with 5G reduces healthcare device costs by offloading tasks to the network edge, extending battery life, and minimizing the need for frequent recharging or replacements, resulting in long-term cost savings [47].

FUTURE SCOPE

The future potential of combining 5G technology with edge computing in the healthcare sector appears highly promising. This synergy holds the capacity to transform healthcare by facilitating instantaneous data processing, minimizing delays, and enhancing the overall effectiveness of healthcare services [47, 48]. Through the amalgamation of technologies like the IoT, AI, and cloud computing, 5G-powered smart healthcare can offer sophisticated and individualized healthcare services [48].

Incorporating edge servers within 5G cellular networks supports widespread mobile computing and enables the provision of healthcare services at the network's edge [49]. A significant advantage of employing edge computing within 5G lies in its support for intelligent machine learning and context-sensitive applications in edge devices [50]. This capability facilitates real-time analysis of health data, thereby enabling early disease detection and the creation of personalized treatment plans [51, 52]. Moreover, the fusion of edge computing with 5G networks can reduce network latency, which is vital for low-latency healthcare applications [52].

Implementing a security system grounded in zero-trust architecture can be employed to guarantee the safeguarding of healthcare data within 5G smart healthcare, ensuring both security and privacy. This framework offers a robust security structure that shields sensitive healthcare data from unauthorized entry, maintaining the confidentiality and integrity of patient information.

Combining blockchain technology with 5G edge networks can bolster the security and reliability of healthcare data. By enabling secure and transparent data sharing, blockchain ensures the steadfastness and unchangeable nature of healthcare records.

To harness the complete capabilities of edge computing in 5G healthcare, various hurdles must be overcome. These obstacles encompass orchestrating resources, optimizing performance, and effectively managing edge computing platforms [53,54]. It is vital to focus on advancing the systems and tools for edge computing to empower developers in the swift creation and deployment of healthcare-related edge applications.

CONCLUSION

Edge computing with 5G offers several advantages in healthcare, including low latency, improved security and privacy, extended battery life for medical devices, and the ability to deploy services closer to the point of care. These advantages have the potential to revolutionize healthcare delivery and improve patient outcomes. The evolution of 5G networks is accompanied by a paradigm shift in cloud computing, moving computing and storage resources closer to the edge of the network, where data is generated and consumed. This decentralized approach, known as Mobile Edge Computing (MEC), offers several advantages, including reduced latency, improved network performance, and enhanced scalability. However, it also introduces new challenges that require careful consideration and research. Edge computing with 5G has the potential to transform healthcare by enabling low-latency, secure, and efficient applications. By bringing computation closer to the data source, edge computing can improve the quality of healthcare services and enable real-time decision-making. Additionally, edge computing can enhance the security and privacy of healthcare systems, while also optimizing resource utilization and energy efficiency. However, the successful deployment of edge computing in healthcare requires a robust network infrastructure, such as MEC, to support the diverse requirements of 5G-enabled healthcare applications.

A capillary distribution of cloud computing capabilities to the edge of the radio access network is made possible by mobile edge computing. This new paradigm reduces backhaul utilisation and processing at the core network while enabling the execution of context-aware and delay-sensitive applications near end users. To create a diverse resource pool, this article suggests investigating the synergies between linked entities in the MEC network. To showcase the advantages of MEC cooperation in 5G networks, we offer three illustrative use scenarios. To provide an overview of the mobile edge ecosystem's development and standardisation path, technical obstacles and unresolved research questions are emphasised.

REFERENCES

[1]   W. Shi, J. Cao, Q. Zhang, Y. Li, and L. Xu, "Edge computing: Vision and challenges", *IEEE Internet Things J.,* vol. 3, no. 5, pp. 637-646, 2016.
[http://dx.doi.org/10.1109/JIOT.2016.2579198]

[5]   S.S. Saranya, and N.S. Fatima, "Exploration on IoT based Edge cloud computing techniques for improving the patient information management system", *2022 International Conference on Edge Computing and Applications (ICECAA),* pp. 1-5, 2022.
[http://dx.doi.org/10.1109/ICECAA55415.2022.9936532]

[2]   L. Chettri, and R. Bera, "A comprehensive survey on internet of things (IoT) toward 5g wireless systems", *IEEE Internet Things J.,* vol. 7, no. 1, pp. 16-32, 2020.
[http://dx.doi.org/10.1109/JIOT.2019.2948888]

[3]   A. Ahad, "A comprehensive review on 5g-based smart healthcare network security: Taxonomy, issues, solutions and future research directions", *Array,* p. 100290, 2023.

[4]   H.N. Qureshi, M. Manalastas, A. Ijaz, A. Imran, Y. Liu, and M.O. Al Kalaa, "Communication requirements in 5g-enabled healthcare applications: review and considerations", *Healthcare (Basel),* vol. 10, no. 2, p. 293, 2022.
[http://dx.doi.org/10.3390/healthcare10020293] [PMID: 35206907]

[5]   M. Hartmann, U. S. Hashmi, and A. Imran, "Edge computing in smart health care systems: Review, challenges, and research directions", *Transactions on Emerging Telecommunications Technologies 33.3,* p. e3710, 2022.
[http://dx.doi.org/10.1002/ett.3710]

[6]   M.S. Al-Rakhami, A. Gumaei, M. Altaf, M.M. Hassan, B.F. Alkhamees, K. Muhammad, and G. Fortino, "Falldef5: a fall detection framework using 5g-based deep gated recurrent unit networks", *IEEE Access,* vol. 9, pp. 94299-94308, 2021.
[http://dx.doi.org/10.1109/ACCESS.2021.3091838]

[7]   D. Li, "5G and intelligence medicine—how the next generation of wireless technology will reconstruct healthcare?", *Precis. Clin. Med.,* vol. 2, no. 4, pp. 205-208, 2019.
[http://dx.doi.org/10.1093/pcmedi/pbz020] [PMID: 31886033]

[8]   T. Hewa, A. Braeken, M. Ylianttila, and M. Liyanage, "Multi-access edge computing and blockchain-based secure telehealth system connected with 5G and IoT", *GLOBECOM 2020 - 2020 IEEE Global*

Communications Conference, Taipei, Taiwan, pp. 1-6, 2020.
[http://dx.doi.org/10.1109/GLOBECOM42002.2020.9348125]

[9]   S. Hamm, A-C. Schleser, J. Hartig, P. Thomas, S. Zoesch, and C. Bulitta, "5G as enabler for Digital Healthcare", *Curr. Dir. Biomed. Eng.,* vol. 6, no. 3, pp. 1-4, 2020.
[http://dx.doi.org/10.1515/cdbme-2020-3001]

[10]   M. Sankaradas, K. Rao, and S. Chakradhar, "AppSlice: A system for application-centric design of 5G and edge computing applications", *12th International Conference on Network of the Future (NoF),* 2021.
[http://dx.doi.org/10.1109/NoF52522.2021.9609821]

[11]   N. Hassan, K.L.A. Yau, and C. Wu, "Edge computing in 5G: A review", *IEEE Access,* vol. 7, pp. 127276-127289, 2019.
[http://dx.doi.org/10.1109/ACCESS.2019.2938534]

[12]   Q. Liang, P. Shenoy, and D. Irwin, "Ai on the edge: rethinking ai-based IoT applications using specialized edge architectures",
[http://dx.doi.org/10.48550/arxiv.2003.12488]

[13]   D. Liu, H. Liang, X. Zeng, Q. Zhang, Z. Zhang, and M. Li, "Edge computing application, architecture, and challenges in ubiquitous power internet of things", *Front. Energy Res.,* vol. 10, p. 850252, 2022.
[http://dx.doi.org/10.3389/fenrg.2022.850252]

[14]   H. Ju, and L. Liu, "Innovation trend of edge computing technology based on patent perspective", *Wirel. Commun. Mob. Comput.,* vol. 2021, no. 1, p. 2609700, 2021.
[http://dx.doi.org/10.1155/2021/2609700]

[15]   F. Wang, M. Zhang, X. Wang, X. Ma, and J. Liu, "Deep learning for edge computing applications: a state-of-the-art survey", *IEEE Access,* vol. 8, pp. 58322-58336, 2020.
[http://dx.doi.org/10.1109/ACCESS.2020.2982411]

[16]   T.X. Tran, A. Hajisami, P. Pandey, and D. Pompili, "Collaborative mobile edge computing in 5g networks: new paradigms, scenarios, and challenges", *IEEE Commun. Mag.,* vol. 55, no. 4, pp. 54-61, 2017.
[http://dx.doi.org/10.1109/MCOM.2017.1600863]

[17]   Delshi Howsalya Devi, "5g technology in healthcare and wearable devices: A review", *Sensors,* p. 2519, 2023.
[http://dx.doi.org/10.3390/s23052519]

[18]   Q.V. Pham, F. Fang, V.N. Ha, M.J. Piran, M. Le, L.B. Le, W-J. Hwang, and Z. Ding, "A survey of multi-access edge computing in 5g and beyond: fundamentals, technology integration, and state-of-the-art", *IEEE Access,* vol. 8, pp. 116974-117017, 2020.
[http://dx.doi.org/10.1109/ACCESS.2020.3001277]

[19]   X. Lin, J. Wu, A.K. Bashir, W. Yang, A. Singh, and A.A. AlZubi, "Fairhealth: long-term proportional fairness-driven 5g edge healthcare in internet of medical things", *IEEE Trans. Industr. Inform.,* vol. 18, no. 12, pp. 8905-8915, 2022.
[http://dx.doi.org/10.1109/TII.2022.3183000]

[20]   A. Moglia, K. Georgiou, B. Marinov, E. Georgiou, R.N. Berchiolli, R.M. Satava, and A. Cuschieri, "5g in healthcare: from covid-19 to future challenges", *IEEE J. Biomed. Health Inform.,* vol. 26, no. 8, pp. 4187-4196, 2022.
[http://dx.doi.org/10.1109/JBHI.2022.3181205] [PMID: 35675255]

[21]   Z. Hong, N. Li, D. Li, J. Li, B. Li, W. Xiong, L. Lu, W. Li, and D. Zhou, "Telemedicine during the covid-19 pandemic: experiences from western china", *J. Med. Internet Res.,* vol. 22, no. 5, p. e19577, 2020.
[http://dx.doi.org/10.2196/19577] [PMID: 32349962]

[22]   Chen, B., Qiao, S., Zhao, J., Liu, D., Shi, X., Lyu, M. & Zhai, Y, "A security awareness and protection

system for 5g smart healthcare based on zero-trust architecture", *Internet of Things Journal,* vol. 8, no. 13, pp. 10248-10263, 2021.
[http://dx.doi.org/10.1109/JIOT.2020.3041042]

[23]   G.C. Kagadis, C. Kloukinas, K. Moore, J. Philbin, P. Papadimitroulas, C. Alexakos, P.G. Nagy, D. Visvikis, and W.R. Hendee, "Cloud computing in medical imaging", *Med. Phys.,* vol. 40, no. 7, p. 070901, 2013.
[http://dx.doi.org/10.1118/1.4811272] [PMID: 23822402]

[24]   Enjie Liu, Youbing Zhao, and Abimbola Efunogbon, "Boosting smarter digital health care with 5G and beyond networks", *ITU Journal on Future and Evolving Technologies,* vol. 4, no. 1, pp. 157-165, 2023.
[http://dx.doi.org/10.52953/GJNN6958]

[25]   J. Harmatos, and M. Maliosz, "Architecture integration of 5g networks and time-sensitive networking with edge computing for smart manufacturing", *Electronics (Basel),* vol. 10, no. 24, p. 3085, 2021.
[http://dx.doi.org/10.3390/electronics10243085]

[26]   K. Pandav, A.G. Te, N. Tomer, S.S. Nair, and A.K. Tewari, "Leveraging 5G technology for robotic surgery and cancer care", *Cancer Rep.,* vol. 5, no. 8, p. e1595, 2022.
[http://dx.doi.org/10.1002/cnr2.1595] [PMID: 35266317]

[27]   M. McClellan, C. Cervelló-Pastor, and S. Sallent, "Deep learning at the mobile edge: opportunities for 5g networks", *Applied Sciences,* vol. 10, no. 14, p. 4735, 2020.
[http://dx.doi.org/10.3390/app10144735]

[29]   T. Lončar-Turukalo, E. Zdravevski, J. Machado da Silva, I. Chouvarda, and V. Trajkovik, "Literature on wearable technology for connected health: scoping review of research trends, advances, and barriers", *J. Med. Internet Res.,* vol. 21, no. 9, p. e14017, 2019.
[http://dx.doi.org/10.2196/14017] [PMID: 31489843]

[30]   E. Covi, E. Donati, X. Liang, D. Kappel, H. Heidari, M. Payvand, and W. Wang, "Adaptive extreme edge computing for wearable devices", *Front. Neurosci.,* vol. 15, p. 611300, 2021.
[http://dx.doi.org/10.3389/fnins.2021.611300] [PMID: 34045939]

[31]   Shah, Syed D., *et al,* "Sdn enhanced multi-access edge computing (mec) for e2e mobility and qos management", *IEEE Access,* vol. 8, pp. 77459-77469, 2020.
[http://dx.doi.org/10.1109/ACCESS.2020.2990292]

[32]   A. Jain, "Enhanced handover signaling through integrated mme-sdn controller solution", *87th Vehicular Technology Conference (VTC Spring),* 2018.
[http://dx.doi.org/10.1109/VTCSpring.2018.8417719]

[33]   A. Ateya, A. Muthanna, I. Gudkova, A. Abuarqoub, A. Vybornova, and A. Koucheryavy, "Development of intelligent core network for tactile internet and future smart systems", *Journal of Sensor and Actuator Networks,* vol. 7, no. 1, p. 1, 2018.
[http://dx.doi.org/10.3390/jsan7010001]

[34]   J. Li, J. Cai, F. Khan, A.U. Rehman, V. Balasubramaniam, J. Sun, and P. Venu, "A secured framework for sdn-based edge computing in IOT-enabled healthcare system", *IEEE Access,* vol. 8, pp. 135479-135490, 2020.
[http://dx.doi.org/10.1109/ACCESS.2020.3011503]

[35]   K. Mohiuddin, M.N. Miladi, M. Ali Khan, M.A. Khaleel, S. Ali Khan, S. Shahwar, O.A. Nasr, and M. Aminul Islam, "Mobile learning new trends in emerging computing paradigms: an analytical approach seeking performance efficiency", *Wirel. Commun. Mob. Comput.,* vol. 2022, pp. 1-17, 2022.
[http://dx.doi.org/10.1155/2022/6151168]

[36]   S. Wijethilaka, and M. Liyanage, "Survey on network slicing for internet of things realization in 5g networks". IEEE Communications Surveys &Amp", *IEEE Commun. Surv. Tutor.,* vol. 23, no. 2, pp. 957-994, 2021.
[http://dx.doi.org/10.1109/COMST.2021.3067807]

[37]   M. Sankaradas, K. Rao, and S. Chakradhar, "AppSlice: A system for application-centric design of 5G and edge computing applications", *12ᵗʰ International Conference on Network of the Future (NoF),* 2021.
[http://dx.doi.org/10.1109/NoF52522.2021.9609821]

[38]   Y. Mao, C. You, J. Zhang, K. Huang, and K.B. Letaief, "A survey on mobile edge computing: the communication perspective". IEEE Communications Surveys &Amp", *IEEE Commun. Surv. Tutor.,* vol. 19, no. 4, pp. 2322-2358, 2017.
[http://dx.doi.org/10.1109/COMST.2017.2745201]

[39]   N. Abbas, Y. Zhang, A. Taherkordi, and T. Skeie, "Mobile edge computing: a survey", *IEEE Internet Things J.,* vol. 5, no. 1, pp. 450-465, 2018.
[http://dx.doi.org/10.1109/JIOT.2017.2750180]

[40]   B. Chen, S. Qiao, J. Zhao, D. Liu, X. Shi, M. Lyu, H. Chen, H. Lu, and Y. Zhai, "A security awareness and protection system for 5g smart healthcare based on zero-trust architecture", *IEEE Internet Things J.,* vol. 8, no. 13, pp. 10248-10263, 2021.
[http://dx.doi.org/10.1109/JIOT.2020.3041042] [PMID: 35783535]

[41]   Md Rahman, "Challenges and prospective of ai and 5g-enabled technologies in emerging applications during the pandemic", 2023.
[http://dx.doi.org/10.5772/intechopen.109450]

[42]   L. Yang, K. Yu, S.X. Yang, C. Chakraborty, Y. Lu, and T. Guo, "An intelligent trust cloud management method for secure clustering in 5g enabled internet of medical things", *IEEE Trans. Industr. Inform.,* vol. 18, no. 12, pp. 8864-8875, 2022.
[http://dx.doi.org/10.1109/TII.2021.3128954]

[43]   T. Sigwele, T. Sigwele, A. Naveed, Y-F. Hu, M. Ali, and M. Susanto, "Energy efficient healthcare monitoring system using 5g task offloading", *Journal of Engineering and Scientific Research,* vol. 1, no. 2, 2019.
[http://dx.doi.org/10.23960/jesr.v1i2.12]

[44]   X. Tang, L. Zhao, J. Chong, Z. You, L. Zhu, H. Ren, Y. Shang, Y. Han, and G. Li, "5G-based smart healthcare system designing and field trial in hospitals", *IET Commun.,* vol. 16, no. 1, pp. 1-13, 2022.
[http://dx.doi.org/10.1049/cmu2.12300]

[45]   Z. Ning, P. Dong, X. Wang, X. Hu, L. Guo, B. Hu, Y. Guo, T. Qiu, and R.Y.K. Kwok, "Mobile edge computing enabled 5G health monitoring for Internet of medical things: A decentralized game theoretic approach", *IEEE J. Sel. Areas Comm.,* vol. 39, no. 2, pp. 463-478, 2021.
[http://dx.doi.org/10.1109/JSAC.2020.3020645]

[46]   Bo Li, "Deployment of edge servers in 5G cellular networks", *Transactions on Emerging Telecommunications Technologies,* 2022.
[http://dx.doi.org/10.1002/ett.3937]

[47]   Lubna Nadeem, "Integration of d2d, network slicing, and mec in 5g cellular networks: survey and challenges", *Ieee Access,* vol. 9, pp. 37590-37612, 2021.
[http://dx.doi.org/10.1109/ACCESS.2021.3063104]

[48]   D. Loghin, S. Cai, G. Chen, T.T.A. Dinh, F. Fan, Q. Lin, J. Ng, B.C. Ooi, X. Sun, Q-T. Ta, W. Wang, X. Xiao, Y. Yang, M. Zhang, and Z. Zhang, "The disruptions of 5g on data-driven technologies and applications", *IEEE Trans. Knowl. Data Eng.,* vol. 32, no. 6, pp. 1179-1198, 2020.
[http://dx.doi.org/10.1109/TKDE.2020.2967670]

[49]   B.C. Ooi, G. Chen, D. Loghin, W. Wang, and M. Zhang, "5g: agent for further digital disruptive transformations", 2019.

[50]   S. Rahmadika, M. Firdaus, S. Jang, and K-H. Rhee, "Blockchain-enabled 5g edge networks and beyond: an intelligent cross-silo federated learning approach", *Secur. Commun. Netw.,* vol. 2021, pp. 1-14, 2021.

[http://dx.doi.org/10.1155/2021/5550153]

[51] I. Dimolitsas, "A delay-aware approach for distributed embedding towards cross-slice communication", *International Mediterranean Conference on Communications and Networking (MeditCom),* 2022.
[http://dx.doi.org/10.1109/MeditCom55741.2022.9928746]

[52] F. Liu, G. Tang, Y. Li, Z. Cai, X. Zhang, and T. Zhou, "A survey on edge computing systems and tools", *Proc. IEEE,* vol. 107, no. 8, pp. 1537-1562, 2019.
[http://dx.doi.org/10.1109/JPROC.2019.2920341]

CHAPTER 7

Big Data Analytics and Machine Learning for Secure and Flexible Mobile Service towards Smart Utilities

Devasis Pradhan[1,*], **Tarique Akhtar**[2] and **Amit Kumar Sahoo**[3]

[1] *Department of Electronics & Communication Engineering, Acharya Institute of Technology, Bangalore, Karnataka, India*

[2] *Data Science Agility, Dubai*

[3] *Lead 1 Workforce Management, UST Global, Bangalore, India*

Abstract: The proliferation of smart utilities has revolutionized the way we manage essential services such as energy, water, and transportation. Mobile technologies play a pivotal role in delivering these services efficiently. However, the sheer volume of data generated by these systems poses significant challenges in terms of security, flexibility, and overall performance. This research explores the synergy of Big Data Analytics and Machine Learning (ML) to address these challenges. We investigate how these technologies can enhance the security of mobile service infrastructures in smart utilities, ensuring the protection of sensitive data and safeguarding against cyber threats. Moreover, we explore the potential of ML algorithms to adapt and optimize mobile service delivery, ensuring flexibility in response to changing demands and environmental conditions. The study leverages real-world data from smart utility deployments, applying advanced analytics techniques to extract valuable insights and patterns. These insights enable the development of proactive security measures and the creation of flexible, adaptive mobile service models. By harnessing the power of Big Data Analytics and ML, we aim to create a foundation for smarter, more secure, and highly responsive mobile services in the context of smart utilities, ultimately contributing to the sustainable development of smart cities and communities.

Keywords: Cybersecurity, Big data analytics, DL, ML, 5G.

INTRODUCTION

In an era where technological advancements continue to reshape our world, the convergence of big data analytics, machine learning, and mobile services stands as a pivotal force, particularly in the realm of smart utilities. As our dependence on efficient and sustainable utility services grows, so does the need for innovative

* **Corresponding author Devasis Pradhan:** Department of Electronics & Communication Engineering, Acharya Institute of Technology, Bangalore, Karnataka, India; E-mail: devasispradhan@acharya.ac.in

Devasis Pradhan, Mangesh M. Ghonge, Nitin S. Goje, Alessandro Bruno and Rajeswari (Eds.)

solutions that enhance their security, flexibility, and overall performance. This introduction sets the stage for exploring the profound impact of big data analytics and machine learning for secure and flexible mobile service towards smart utilities [1, 2].

The Era of Smart Utilities

The dawn of smart utilities marks a transformative shift in how we manage, distribute, and consume essential resources. From electricity and water to waste management, the integration of advanced technologies promises to make these services more intelligent, responsive, and tailored to the needs of a dynamic society. At the heart of this evolution lies the synergy between big data analytics, machine learning, and mobile services [3].

Significance of Mobile Services in Smart Utilities

Mobile services play a crucial role in connecting and optimizing various components of smart utility systems. From monitoring and control to real-time data acquisition, mobile platforms facilitate the seamless operation of utility infrastructure. However, as these services become more integral, the challenges of ensuring their security and flexibility become increasingly complex [4].

Triumvirate of Big Data, Machine Learning, and Mobile Services

Big data analytics serves as the backbone, empowering utilities to extract meaningful insights from vast datasets. Machine learning algorithms, in turn, enable utilities to predict and adapt to changing conditions, optimizing operations and resource allocation. The amalgamation of these technologies with mobile services creates a dynamic ecosystem capable of responding intelligently to the needs of both providers and consumers [5].

Security Imperatives in Smart Utilities

As the connectivity and interdependence of smart utilities intensify, so does the need for robust security measures. Cyber threats loom large, making secure mobile services an imperative. Encryption, authentication, and proactive security measures become essential components in safeguarding critical utility infrastructure [6].

Flexibility for Dynamic Environments

Smart utility operations are inherently dynamic, influenced by factors ranging from weather conditions to user demand. Machine learning algorithms, when integrated with mobile services, enable a level of adaptability that ensures utilities can respond nimbly to changing circumstances, optimizing efficiency and resource utilization. This exploration into big data analytics and machine learning for secure and flexible mobile service towards smart utilities will unfold across chapters, each delving into specific aspects of this intricate ecosystem. From laying the foundations of big data analytics and machine learning to dissecting case studies demonstrating their real-world impact, the following pages will offer insights into the challenges, opportunities, and ethical considerations inherent in this transformative journey [7, 8].

THE FOUNDATIONS OF BIG DATA ANALYTICS

In the landscape of smart utilities, big data refers to the vast volume, velocity, and variety of data generated by various components within the utility infrastructure. These components include sensors, smart meters, control systems, and other IoT devices that continuously produce a torrent of information. Big data in smart utilities encompasses structured and unstructured data, offering insights into the consumption patterns, operational performance, and overall health of utility systems. The three key dimensions of big data – volume, velocity, and variety – are particularly pronounced in the context of smart utilities. The sheer volume of data generated on a real-time basis, the rapid velocity at which it is produced, and the diverse formats it takes present unique challenges and opportunities. Big data in smart utilities becomes a strategic asset when harnessed effectively, providing a comprehensive view of the entire utility ecosystem [8, 9].

Big data analytics plays a pivotal role in transforming raw data into actionable insights, thereby optimizing utility operations and decision-making processes. By leveraging advanced analytics tools, utilities can gain a granular understanding of consumer behavior, equipment performance, and system vulnerabilities. This, in turn, enables utilities to enhance operational efficiency, reduce downtime, and proactively address potential issues. The utilization of big data in smart utilities extends to predictive analytics, allowing for the anticipation of equipment failures or fluctuations in demand. With these predictive capabilities, utilities can implement preventive measures, schedule maintenance more efficiently, and ultimately reduce costs. Additionally, big data-driven decision-making enables utilities to respond dynamically to changing conditions, fostering a more adaptive and resilient infrastructure discussed in Table **1** [9 - 11].

These examples illustrate how big data analytics empowers utilities to extract actionable insights from the massive volumes of data they generate. By harnessing the potential of big data, smart utilities can achieve operational excellence, make informed decisions, and ultimately provide more reliable and efficient services to consumers.

Table 1. Successful applications of big data analytics in the utility sector.

S. No.	Application Area	Important Points
1	Demand Forecasting	• Big data analytics enables utilities to analyze historical consumption patterns, weather data, and other relevant factors to predict future energy demand. • This facilitates proactive resource planning, preventing overloads and ensuring a stable supply.
2	Grid Management	• Smart grids, equipped with sensors and IoT devices, generate vast amounts of data. • Big data analytics processes this information to optimize grid performance, detect anomalies, and enhance overall reliability.
3	Asset Management	• Utilities leverage big data to monitor the health and performance of critical assets such as transformers and turbines. • Predictive maintenance models analyze data to anticipate potential failures, allowing for timely interventions and extending the lifespan of equipment.
4	Consumer Engagement	• Big data analytics helps utilities understand consumer behavior and preferences. • This information enables the development of targeted programs, such as demand-response initiatives, to encourage efficient energy usage among consumers.
5	Water Management	• In water utilities, big data analytics aids in monitoring water quality, detecting leaks, and optimizing distribution. • This ensures the sustainable use of water resources and reduces wastage.

MACHINE LEARNING IN SMART UTILITY MANAGEMENT

Machine learning (ML) represents a paradigm shift in how smart utilities manage and optimize their operations. At its core, machine learning involves the development of algorithms that enable systems to learn from data, identify patterns, and make predictions or decisions without explicit programming. In the context of smart utilities, where large volumes of data are generated continuously, machine learning emerges as a powerful tool to extract meaningful insights, enhance operational efficiency, and improve decision-making processes. Machine learning in smart utilities allows systems to adapt and evolve in response to changing conditions, making them more resilient and capable of addressing complex challenges. From predicting equipment failures to optimizing resource allocation, machine learning algorithms play a crucial role in transforming raw data into actionable intelligence [12, 13].

Overview of Machine Learning Algorithms for Predictive Maintenance and Real-Time Decision-Making

- **Regression Algorithms:** Regression models, such as linear regression or polynomial regression, are used to predict numerical values, making them valuable for forecasting parameters like energy consumption or equipment degradation.
- **Decision Trees:** Decision tree algorithms are effective in classifying data into categories. In smart utilities, decision trees can be employed to categorize different types of equipment failures or to assess the risk of specific events occurring.
- **Random Forests:** Random Forests combine multiple decision trees to improve accuracy and reduce overfitting. They are suitable for complex utility scenarios where various factors influence outcomes.
- **Support Vector Machines (SVM):** SVM is used for classification tasks and can be applied to identify patterns or anomalies in utility data. This is particularly useful in detecting irregularities in grid behavior or equipment performance.
- **Neural Networks:** Neural networks, inspired by the human brain, are adept at handling complex relationships within data. In smart utilities, neural networks can be employed for tasks like load forecasting and fault detection.
- **Clustering Algorithms:** Clustering algorithms, such as K-means, help identify groups or patterns within datasets. In the context of smart utilities, clustering can be used to categorize consumers based on usage patterns or to group similar types of equipment for maintenance planning.

These machine learning algorithms contribute to the predictive maintenance and real-time decision-making capabilities of smart utilities. By analyzing historical data, identifying trends, and adapting to evolving conditions, these algorithms empower utilities to proactively manage their infrastructure, minimize downtime, and enhance overall reliability.

Case Studies Illustrating the Impact of Machine Learning on Smart Utility Management

Case 1: Predictive Maintenance in Power Distribution: A utility company implemented machine learning algorithms to predict potential failures in power distribution infrastructure. By analyzing historical data on equipment performance, the system accurately forecasted maintenance needs, reducing unplanned downtime and maintenance costs.

Case 2: Load Forecasting for Energy Distribution: Machine learning models were employed to forecast energy demand patterns, enabling utilities to optimize energy distribution and allocate resources efficiently. This resulted in improved load balancing and reduced instances of grid congestion.

Case 3: Anomaly Detection in Water Treatment: In a water utility, machine learning algorithms were used to detect anomalies in water quality data. The system identified deviations from normal patterns, allowing for rapid response to potential contamination events and ensuring the delivery of safe water to consumers.

Case 4: Dynamic Pricing Strategies: Machine learning algorithms were utilized to analyze consumer behavior and preferences in a smart utility's electricity consumption patterns. This information was then leveraged to implement dynamic pricing strategies, encouraging consumers to shift their usage during non-peak hours.

These case studies demonstrate the tangible impact of machine learning on smart utility management. By leveraging predictive capabilities and real-time decision-making, utilities can enhance their reliability, efficiency, and sustainability, ultimately providing better services to consumers and contributing to the advancement of smart infrastructure [13 - 15].

ENSURING SECURE MOBILE SERVICES

In the dynamic landscape of smart utilities, the role of mobile services is integral to seamless and efficient operations. However, with the increasing reliance on mobile platforms, the significance of ensuring the security of these services becomes paramount. Secure mobile services not only protect sensitive data but also safeguard critical utility infrastructure, ensuring the reliability and trustworthiness of the entire system. This section delves into the crucial importance of security measures in the realm of smart utilities, emphasizing the need for a robust framework to mitigate potential risks [14].

Parameter plays important role: Encryption, Authentication, and Other Security Measures

Encryption

Encryption serves as a cornerstone of modern cybersecurity, providing a robust defense against unauthorized access, data breaches, and privacy violations. At its core, encryption involves the transformation of readable data (plaintext) into a coded format (ciphertext) using mathematical algorithms and cryptographic keys.

This transformative process ensures that even if unauthorized entities gain access to the encrypted data, it remains incomprehensible without the corresponding decryption key. In the realm of smart utility mobile services, encryption acts as a fundamental safeguard for sensitive information, including user credentials, transaction data, and communication channels. By implementing encryption, utilities bolster the confidentiality and integrity of their data, establishing a secure foundation for mobile interactions within the smart utility ecosystem [17].

The Role of End-to-End Encryption in Securing Data Transmission

End-to-end encryption (E2EE) is a specialized form of encryption that ensures the confidentiality of data throughout its entire journey, from the sender to the recipient. In the context of smart utility mobile services, E2EE plays a crucial role in protecting the privacy of user communications and preventing unauthorized interception or tampering [18, 19].

Key Aspects of End-To-End Encryption in Smart Utility Mobile Services Include

- **User Privacy:** E2EE ensures that only the intended recipient can decipher the encrypted data, preserving the confidentiality of user communications. This is particularly important when transmitting sensitive information related to utility usage, billing, or personal details.
- **Mitigating Intermediary Risks:** By encrypting data end-to-end, the risks associated with intermediaries, such as communication service providers or network administrators, are mitigated. Even if data passes through various points in the communication chain, only the authorized endpoints possess the keys necessary for decryption.
- **Data Integrity:** E2EE not only focuses on confidentiality but also ensures the integrity of transmitted data. Any attempt to tamper with the encrypted information would result in decryption failure, alerting both the sender and recipient to potential security breaches.
- **Secure Collaboration:** For smart utility applications that involve collaborative processes or shared data, E2EE enhances the security of collaborative interactions, allowing users to exchange information without fear of unauthorized access.

Encryption Algorithms and Protocols Employed to Protect Data Integrity and Confidentiality in Smart Utility Mobile Services

- **AES (Advanced Encryption Standard):** Often considered the gold standard in encryption, AES is widely employed in smart utilities for securing data at rest and in transit. Its strength lies in its efficiency, reliability, and resistance to brute-force attacks.

- **RSA (Rivest-Shamir-Adleman):** RSA is a widely used asymmetric encryption algorithm for securing communication channels and facilitating secure key exchange. It plays a crucial role in digital signatures and the establishment of secure connections.
- **TLS (Transport Layer Security):** TLS is a protocol that ensures secure communication over a computer network, commonly employed to safeguard data transmissions between users and smart utility servers. It encrypts the data, providing a secure channel for information exchange.
- **IPsec (Internet Protocol Security):** IPsec is utilized to secure communication at the network layer, ensuring the confidentiality and integrity of data as it traverses networks. In smart utilities, IPsec can be instrumental in securing data transmissions between different utility components.
- **VPN (Virtual Private Network):** While not strictly an encryption algorithm, VPNs use encryption protocols to create secure, encrypted tunnels for data transmission. In smart utility mobile services, VPNs can enhance the security of remote access and communication.

By employing these encryption algorithms and protocols, smart utilities can establish a robust defense against potential security threats, ensuring the confidentiality, integrity, and privacy of data transmitted through mobile services. This comprehensive approach to encryption contributes to the overall security posture of smart utility systems, instilling confidence in users and stakeholders alike [20 - 22].

Authentication

Authentication stands as the first line of defense against unauthorized access in the realm of smart utility mobile services. In an era where cyber threats continually evolve, ensuring robust authentication mechanisms is paramount to safeguarding sensitive information, user accounts, and critical utility infrastructure. Strong authentication serves as a barrier, preventing unauthorized individuals or entities from gaining access to privileged systems or data [23, 24].

Key Points Highlighting the Importance of Strong Authentication Include

- **User Identity Protection:** Strong authentication ensures that users are who they claim to be, mitigating the risk of identity theft or impersonation. This is particularly critical in the context of smart utility mobile services, where user identities are closely linked to personal and utility-related data.

- **Data Confidentiality:** Unauthorized access can lead to the exposure of sensitive information, such as utility bills, consumption patterns, and personal details. Strong authentication safeguards the confidentiality of this data, preventing unauthorized parties from extracting valuable insights.
- **Preventing Unauthorized Transactions:** With the integration of mobile services for utility payments and transactions, strong authentication acts as a barrier against fraudulent activities. It ensures that only authorized users can initiate and approve financial transactions.
- **Protecting Critical Infrastructure:** In smart utility systems, unauthorized access could potentially compromise critical infrastructure, leading to service disruptions or even physical damage. Strong authentication mechanisms play a crucial role in preventing malicious actors from gaining control over utility components.

Multi-Factor Authentication as an Additional Layer of Security for Mobile Services

Multi-factor authentication (MFA) adds an extra layer of security by requiring users to provide multiple forms of identification before granting access [25]. In the context of smart utility mobile services, MFA serves as a robust defense against unauthorized access attempts. It typically involves a combination of the following factors:

- **Knowledge Factor:** This includes traditional passwords or personal identification numbers (PINs) that users must enter.
- **Possession Factor:** Users may be required to possess a physical token (such as a smart card or USB security key) or receive a one-time code *via* SMS or email.
- **Biometric Factor:** Biometric authentication methods, such as fingerprint scans or facial recognition, add a unique biological element to the authentication process.

Benefits of Multi-factor Authentication in Smart Utility Mobile Services

- **Enhanced Security:** MFA significantly strengthens the authentication process, requiring attackers to compromise multiple factors to gain unauthorized access.
- **Reduced Vulnerability to Password Attacks:** MFA mitigates the risks associated with password-related vulnerabilities, as even if a password is compromised, an additional authentication factor remains.
- **Compliance Requirements:** In many regions, regulations and standards mandate the use of multi-factor authentication for certain types of sensitive data and transactions.

Biometric Authentication and its Application in Enhancing the Security of Utility-related Mobile Applications

Biometric authentication leverages unique physical or behavioral traits of individuals, such as fingerprints, facial features, or voice patterns, for user identification. In the context of smart utility mobile applications, biometric authentication provides an additional layer of security while enhancing user convenience [25, 26]. Key considerations for biometric authentication in smart utility mobile applications are as follows:

- **Unique Identifiers:** Biometric traits are unique to individuals, providing a highly secure means of authentication. This uniqueness makes it challenging for unauthorized users to replicate or forge.
- **User Experience:** Biometric authentication offers a seamless and user-friendly experience, eliminating the need for users to remember and input complex passwords.
- **Enhanced Security:** Biometric methods add an extra layer of security beyond traditional authentication, making it more difficult for malicious actors to gain unauthorized access.
- **Use Cases in Smart Utilities:** Biometric authentication can be applied to access sensitive utility data, make secure payments, or control smart home devices linked to utility services.
- **Continuous Advancements:** Ongoing advancements in biometric technology, such as behavioral biometrics, further enhance the security and accuracy of user identification.

By incorporating strong authentication mechanisms, multi-factor authentication, and biometric authentication into smart utility mobile services, organizations can establish a robust defense against unauthorized access. These measures not only protect user accounts and sensitive data but also contribute to the overall resilience and security of smart utility systems in the face of evolving cyber threats.

Network Security

Secure communication protocols play a pivotal role in safeguarding data during transmission, especially in the context of smart utility mobile services where sensitive information is exchanged. Two key protocols that significantly enhance network security are HTTPS (Hypertext Transfer Protocol Secure) and VPNs (Virtual Private Networks). Table **2** discusses the importance of HTTPS and VPNs [27 - 29].

Table 2. Importance of HTTPs and VPNs.

SL. No.	Protocols	Important Factor
1	HTTPS	• Encryption of Data in Transit: HTTPS ensures the encryption of data between a user's device and the server, protecting it from eavesdropping or interception. • Authentication and Data Integrity: By using digital certificates, HTTPS verifies the authenticity of the server and ensures the integrity of transmitted data. • Secure Web Applications: Smart utility mobile services often rely on web applications. Implementing HTTPS ensures that data exchanged between users and these applications remains confidential and secure.
2	VPNs	• Securing Data Over Public Networks: VPNs create secure, encrypted tunnels over public networks, ensuring the confidentiality and integrity of data transmitted between mobile devices and utility servers. • Remote Access Security: For smart utility applications that involve remote access, VPNs provide a secure method for users to connect to utility networks from various locations. • Protection against Man-in-the-Middle Attacks: VPNs protect against man-in-the-middle attacks by encrypting the communication channel, preventing unauthorized interception and tampering.

Implementation of Firewalls and Intrusion Detection/Prevention Systems to Protect Mobile Services

Firewalls and intrusion detection/prevention systems (IDS/IPS) are essential components of network security, acting as barriers against unauthorized access,cyber threats, and malicious activities. Table **3** discusses important factors regarding Fiewall and IDS/IPS.

Table 3. Important factor based on firewall and IDS/IPS.

SL.No.	Component of Network Security	Factors	Remarks
1	Firewall	Access Control	Firewalls regulate incoming and outgoing network traffic, enforcing access control policies and preventing unauthorized entities from accessing the network.
		Protection Against Cybe	By inspecting and filtering network traffic, firewalls mitigate the risks associated with malware, ransomware, and other cyber threats.
		Application Layer Security	Next-generation firewalls provide deep packet inspection at the application layer, offering granular control over data flows and preventing attacks targeting specific applications.

(Table 3) cont.....

SL.No.	Component of Network Security	Factors	Remarks
2	IDS/IPS	Real-time Monitoring	IDS/IPS continuously monitors network and system activities in real-time, identifying patterns indicative of potential security incidents or attacks.
		Immediate Response	In the event of detected threats, IDS/IPS can take immediate action, blocking or mitigating the impact of the threat.
		Signature-based and Anomaly-based Detection	IDS/IPS use signature-based detection for known threats and anomaly-based detection to identify deviations from normal network behavior.

Securing wireless networks and mobile data transmissions is crucial for maintaining the integrity and confidentiality of smart utility mobile services, especially given the prevalence of wireless communication in modern scenarios.

- **Use of WPA3 (Wi-Fi Protected Access 3):** WPA3 is the latest Wi-Fi security protocol, offering stronger encryption and protection against brute-force attacks. Implementing WPA3 enhances the security of wireless networks.
- **Network Segmentation:** Segmenting the network isolates different components, preventing unauthorized lateral movement. This is particularly important in smart utility environments where various systems may have different security requirements.
- **Mobile Device Management (MDM):** MDM solutions enable organizations to enforce security policies on mobile devices accessing utility networks. This includes measures such as device encryption, remote wiping, and application control.

Security Best Practices for Mobile Data Transmissions

- Encourage the use of secure channels, such as VPNs, for remote access.
- Regularly update and patch mobile devices and applications to address vulnerabilities.
- Implement strong authentication mechanisms, including biometric authentication for mobile access.

By incorporating these measures into the network architecture of smart utility mobile services, organizations can establish a robust security framework. This not only protects against potential threats but also ensures the reliability and integrity of data transmitted across mobile networks in the smart utility ecosystem.

Device Security

End-user devices, such as smartphones and tablets, serve as the primary interface for consumers to access and interact with smart utility services. The security of these devices is of paramount importance as they act as gateways to sensitive information and functionalities related to utility consumption, payments, and personal data. Table **4** portrays Securing end-user devices is critical for several reasons.

Table 4. Securing end-user devices is critical for several reasons.

Sl. No.	Critical Scenario	Remarks
1	Protection of Sensitive Information	• Smart utility applications often handle sensitive information, including personal details, billing information, and consumption patterns. • Securing end-user devices ensures the confidentiality of this data.
2	Prevention of Unauthorized Access	• Unsecured devices are susceptible to unauthorized access, putting users at risk of identity theft or unauthorized use of utility services. • Implementing robust security measures safeguards against such risks.
3	Ensuring the Integrity of Utility Transactions	• Secure end-user devices contribute to the integrity of utility transactions, preventing tampering or unauthorized alterations of data related to payments, usage patterns, and other critical information.
4	Protecting Against Malware and Cyber Threats	• End-user devices are susceptible to malware, ransomware, and other cyber threats. • Securing these devices with up-to-date antivirus software and security patches is essential to prevent malicious activities.
5	Maintaining User Trust	• Users need to trust that their interactions with smart utility services through their devices are secure. • Device security measures contribute to building and maintaining this trust, ensuring a positive user experience.

Mobile Device Management (MDM) Solutions for Enforcing Security Policies

Mobile Device Management (MDM) solutions are integral to enforcing security policies on end-user devices, especially in the context of smart utility services. MDM solutions provide organizations with the tools to manage, monitor, and secure mobile devices accessing utility services. Key aspects of MDM solutions include:

• **Enforcement of Security Policies:** MDM solutions enable organizations to define and enforce security policies on end-user devices. This includes requirements for strong passwords, encryption, and the installation of security updates.

- **Remote Device Management:** In the event of a lost or stolen device, MDM solutions allow for remote actions, such as locating the device, locking it, or initiating a remote wipe to protect sensitive data.
- **Application Control:** MDM solutions can control which applications are allowed on end-user devices. This ensures that only authorized and secure applications are used to access utility services.
- **Monitoring Device Health:** MDM solutions provide continuous monitoring of device health, detecting potential security threats or vulnerabilities. Alerts and actions can be triggered in response to security incidents.
- **Compliance Tracking:** Organizations can ensure that end-user devices comply with security standards and policies through MDM solutions. Non-compliant devices can be identified and remediated to maintain a secure environment.

Application of Secure Boot Processes and Device Encryption to Enhance Overall Device Security

- **Secure Boot Processes:** Secure boot processes ensure that only trusted and signed software components are allowed to run on a device during the boot-up sequence. This prevents the execution of unauthorized or tampered firmware, protecting the integrity of the device's operating system.
- **Device Encryption:** Device encryption is crucial for protecting the data stored on end-user devices. It involves encoding the data in a way that can only be accessed or decrypted with the appropriate authentication credentials. This safeguards user information, utility-related data, and any sensitive files stored on the device.
- **Protection against Physical Attacks:** Secure boot processes and device encryption add a layer of defense against physical attacks on the device. Even if an attacker gains physical access to the device, the encrypted data remains inaccessible without the proper credentials.
- **Compliance with Privacy Regulations:** Many privacy regulations and standards require the use of encryption to protect user data. Implementing device encryption ensures that organizations comply with these regulations, avoiding potential legal and reputational risks.

By implementing these measures—securing end-user devices, leveraging MDM solutions, and applying secure boot processes and device encryption—organizations can establish a robust defense against potential security threats and vulnerabilities in the context of smart utility services. This holistic approach enhances overall device security, protecting both users and the integrity of utility-related data and transactions [25, 27].

Security Audits and Monitoring

Regular security audits are a proactive and essential measure to assess and strengthen the security posture of smart utility mobile services. These audits involve systematic evaluations of security policies, procedures, systems, and infrastructure to identify vulnerabilities and potential risks. The basic aspects of security audits include:

- **Risk Assessment:** Identify and assess potential risks and vulnerabilities in the smart utility mobile services ecosystem. This includes analyzing the application, network, and system vulnerabilities that could be exploited by attackers.
- **Compliance Verification:** Ensure that security practices align with industry regulations, standards, and organizational policies. Regular audits help verify compliance and address any deviations from established security requirements.
- **Penetration Testing:** Conduct penetration tests to simulate real-world cyberattacks and assess the resilience of the mobile services. Identify and remediate vulnerabilities before malicious actors can exploit them.
- **Configuration Reviews:** Review and validate the configurations of mobile applications, servers, and network devices to ensure they adhere to security best practices. Incorrect configurations can introduce vulnerabilities that might be exploited.
- **User Access Audits:** Evaluate user access controls and permissions to prevent unauthorized access. Regular audits help identify and revoke unnecessary or excessive user privileges, reducing the risk of insider threats.
- **Infrastructure Assessments:** Assess the security of the underlying infrastructure, including servers, databases, and cloud services. Identify and patch vulnerabilities to protect against potential exploits.

Continuous Monitoring of Mobile Service Activities for Anomalous Behavior

Continuous monitoring is crucial for detecting anomalous behavior and potential security incidents in real time. This ongoing surveillance allows organizations to identify and respond promptly to security threats. The basic components of continuous monitoring include:

- **Behavioral Analytics:** Utilize behavioral analytics to establish baseline behavior for users and devices accessing smart utility mobile services. Deviations from established patterns may indicate potential security incidents.
- **Network Traffic Analysis:** Monitor network traffic for unusual patterns or spikes in activity. Anomalous network behavior could be indicative of a security threat or a compromised device.
- **User Activity Monitoring:** Track user activities within smart utility applications to detect any unusual or unauthorized actions. This includes

monitoring login attempts, data access, and transactions for signs of suspicious behavior.

- **Endpoint Security Monitoring:** Employ endpoint security solutions to monitor the security status of end-user devices. Detect and respond to security events on individual devices accessing utility services.
- **Log Analysis:** Analyze logs generated by mobile applications, servers, and network devices for indicators of compromise. Correlate information from various sources to identify potential security incidents.
- **Threat Intelligence Integration:** Integrate threat intelligence feeds to stay informed about emerging threats. Continuous monitoring with threat intelligence helps organizations anticipate and mitigate potential risks.

Continuous monitoring provides a proactive defense against evolving cyber threats by enabling rapid detection and response to anomalous behavior. This approach is crucial for maintaining the security and integrity of smart utility mobile services in real time [26, 28].

Incident Response Planning to Address Security Breaches Promptly

Despite proactive measures, security incidents may still occur. Having a well-defined incident response plan is crucial for minimizing the impact of security breaches and responding promptly to contain and remediate the situation [29]. The basic elements of incident response planning include:

- **Incident Detection and Reporting:** Establish clear procedures for detecting and reporting security incidents. Define criteria that indicate a potential security breach and mechanisms for users and administrators to report incidents.
- **Incident Response Team:** Form an incident response team with clearly defined roles and responsibilities. This team should include individuals with expertise in cybersecurity, forensics, legal, and communication to address different aspects of an incident.
- **Communication Protocols:** Define communication protocols for internal and external stakeholders. Establish clear lines of communication to keep relevant parties informed about the incident, including employees, customers, regulators, and law enforcement if necessary.
- **Containment and Eradication:** Develop procedures for containing and eradicating the security incident. This may involve isolating affected systems, removing malicious components, and implementing corrective measures to prevent a recurrence.

- **Forensic Analysis:** Conduct a thorough forensic analysis to understand the scope and impact of the incident. Identify the root cause, assess the extent of data compromise, and gather evidence for potential legal or regulatory requirements.
- **Legal and Regulatory Compliance:** Ensure that incident response actions comply with legal and regulatory requirements. Promptly report incidents to relevant authorities, if required, and follow established procedures for legal and regulatory compliance.
- **Post-Incident Review:** Conduct a post-incident review to evaluate the effectiveness of the incident response plan. Identify areas for improvement and update the plan based on lessons learned from the incident.

Having a well-prepared incident response plan is crucial for minimizing the duration and impact of security incidents on smart utility mobile services. It ensures a coordinated and efficient response, helping organizations recover quickly and implement measures to prevent future incidents.

INTEGRATING BIG DATA ANALYTICS AND MACHINE LEARNING IN MOBILE SERVICES

In the realm of smart utilities, integrating big data analytics and machine learning with mobile services can revolutionize operations, decision-making processes, and user experiences. This synergy enhances the efficiency and flexibility of utility services by leveraging the power of data-driven insights and predictive capabilities [26].

Data Collection and Processing

- **Big Data Analytics:** Big data analytics facilitates the collection, storage, and processing of vast amounts of data generated by smart utility systems. This includes data from smart meters, sensors, user interactions, and other sources.
- **Machine Learning:** Machine learning algorithms can analyze and interpret this diverse and often complex data, identifying patterns, trends, and anomalies. This enables utilities to gain valuable insights into user behavior, energy consumption patterns, and system performance.

Predictive Maintenance

- **Big Data Analytics:** Predictive maintenance models can be developed using historical data to anticipate equipment failures or maintenance needs. Big data analytics processes large datasets to identify patterns indicative of potential issues.

- **Machine Learning:** Machine learning algorithms contribute by continuously learning from real-time and historical data, improving the accuracy of predictions over time. This proactive approach minimizes downtime and enhances the reliability of utility services.

Energy Consumption Forecasting

- **Big Data Analytics:** Big data analytics enables the analysis of historical energy consumption patterns, considering factors such as weather, time of day, and user behavior.
- **Machine Learning:** Machine learning algorithms, particularly regression models and neural networks, can forecast future energy demands. This aids utilities in optimizing energy distribution, resource allocation, and load balancing.

User Behavior Analysis

- **Big Data Analytics:** Big data analytics processes user interaction data from mobile services, such as app usage patterns, preferences, and feedback.
- **Machine Learning:** Machine learning algorithms analyze this data to understand user behavior, enabling utilities to personalize services, offer targeted recommendations, and improve user engagement.

Dynamic Pricing Strategies

- **Big Data Analytics:** Analysis of consumption patterns, market trends, and external factors.
- **Machine Learning:** Algorithms can predict optimal pricing strategies based on user behavior, helping utilities implement dynamic pricing models that incentivize energy consumption during off-peak hours.

ADAPTIVE MOBILE SERVICES FOR DYNAMIC UTILITY ENVIRONMENTS

Utility operations are inherently dynamic, influenced by factors such as fluctuating energy demand, varying weather conditions, and evolving user behaviors. To effectively navigate this complexity, adaptive mobile services play a crucial role in ensuring that smart utilities remain responsive, efficient, and user-centric. Table **5** discuss about various model. The adaptability of these services is paramount for addressing the dynamic nature of utility environments [12, 13, 25].

Table 5. Exploration of machine learning models designed to adapt mobile services.

Sl. No.	Features	Remarks	Example
1	Reinforcement Learning for Dynamic Optimization	Reinforcement learning models can adapt mobile services by learning optimal strategies through continuous interaction with dynamic utility environments.	An adaptive energy management app using reinforcement learning to optimize home energy consumption patterns based on changing electricity prices and grid conditions.
2	Predictive Analytics for Anticipatory Adaptation	Machine learning models, such as predictive analytics, can forecast changes in utility operations, allowing mobile services to proactively adapt.	An adaptive billing system using predictive analytics to anticipate peak demand periods and provide users with cost-saving recommendations.
3	User Behavior Analysis for Personalization	Machine learning algorithms analyzing user behavior can adapt mobile services by personalizing interfaces, recommendations, and notifications.	A smart thermostat app using machine learning to understand user preferences and adjust heating and cooling settings automatically.
4	Context-Aware Recommendation Systems	Context-aware recommendation systems leverage machine learning to adapt mobile services based on user context, location, and real-time conditions.	A utility app recommends energy-efficient appliances when users are in proximity to an appliance store, considering both user preferences and local promotions.
5	Anomaly Detection for Security and Reliability	Machine learning models can detect anomalies in utility operations, enhancing the security and reliability of mobile services by adapting to potential threats.	An adaptive security system using anomaly detection to identify abnormal patterns in user logins and transactions, triggering immediate response measures.

Dynamic Nature of Utility Operations

- Utility operations experience constant fluctuations influenced by factors like peak energy demand, renewable energy variability, and real-time grid conditions.
- Weather patterns, user behaviors, and external events contribute to the dynamic nature of utility environments, necessitating flexible and adaptive solutions.

Necessity for Adaptive Mobile Services

- **Real-time Responsiveness:** Adaptive mobile services can dynamically respond to changes in utility operations in real-time, providing users with up-to-date information and optimizing service delivery.

- **User Experience:** In a dynamic environment, user needs and preferences may change rapidly. Adaptive mobile services ensure that user interfaces, features, and recommendations evolve to meet changing expectations.

The case studies in Table **6** demonstrate tangible benefits of adaptive mobile services in addressing the dynamic challenges of smart utilities. By leveraging machine learning models, utilities can optimize operations, enhance user experiences, and contribute to a more resilient and responsive energy ecosystem.

Table 6. Case studies demonstrating the benefits of adaptive mobile services in smart utilities.

Sl. No.	Features	Challenges	Adaptive Solution	Benefits
1	Dynamic Pricing Optimization	A utility faced fluctuating electricity prices and changing user consumption patterns.	Implemented a dynamic pricing optimization mobile service using machine learning to adjust pricing in real-time based on demand and supply conditions.	Reduced peak demand, optimized user bills, and improved grid stability.
2	Proactive Demand Response	A utility needed to manage peak demand and prevent grid overloads	Deployed a mobile app using reinforcement learning to predict peak demand periods and incentivize users to shift energy consumption through timely notifications and rewards.	Benefits: Improved grid reliability, reduced energy costs, and enhanced user engagement.
3	Personalized Energy Management	A utility sought to engage users in energy conservation efforts	Introduced a personalized energy management app using user behavior analysis to provide customized tips, alerts, and real-time insights.	Increased user awareness, enhanced energy efficiency, and positive feedback from users.
4	Adaptive Billing and Notifications	A utility faced challenges in keeping users informed about billing and consumption changes	Implemented an adaptive billing and notification system using predictive analytics to anticipate billing cycles and notify users in advance.	Improved user satisfaction, reduced billing-related queries, and increased billing accuracy.
5	Grid Anomaly Detection and Response	A utility needed to enhance grid security and respond promptly to anomalies	Integrated an adaptive mobile service with machine learning models for anomaly detection, triggering automated responses and alerts for potential security threats.	Enhanced grid security, minimized downtime, and improved incident response capabilities.

REAL-WORLD CASE STUDIES IN SMART UTILITY OPTIMIZATION

The case studies illustrate the transformative impact of big data analytics and machine learning in optimizing utility operations. Key takeaways include the importance of proactive maintenance, user incentives for demand response, and the critical role of accurate and timely data in decision-making. Challenges such as data integration and user education can be addressed through standardized processes and transparent communication.

Case Study 1

Predictive Maintenance in Power Distribution Systems

Background

A major power utility company implemented a predictive maintenance system leveraging big data analytics and machine learning to enhance the reliability and efficiency of its power distribution infrastructure.

Implementation

- **Data Collection:** Collected real-time data from sensors embedded in power transformers, circuit breakers, and other critical components.
- **Predictive Analytics:** Utilized big data analytics to process and analyze large datasets, identifying patterns and anomalies.
- **Machine Learning Models:** Trained machine learning models to predict equipment failures based on historical data and real-time inputs.
- **Integration with Maintenance Scheduling:** Integrated the predictive maintenance system with the utility's maintenance scheduling software to optimize repair and replacement activities.

Results

- **Reduced Downtime:** Predictive maintenance significantly reduced unplanned downtime by addressing potential issues before they led to failures.
- **Cost Savings:** Efficient maintenance scheduling led to cost savings by minimizing the need for emergency repairs and optimizing resource allocation.
- **Improved Equipment Lifespan:** Proactive identification of potential failures extended the lifespan of critical equipment.

Key Takeaways

- **Proactive Maintenance:** Predictive maintenance, enabled by big data analytics and machine learning, shifts maintenance practices from reactive to proactive, improving overall system reliability.
- **Data Quality Matters:** The success of predictive maintenance relies on the quality of data collected. Accurate and timely data is crucial for training accurate machine learning models.

Challenges and Solutions

- **Data Integration:** Integrating data from diverse sources was a challenge. The solution involved adopting standardized data formats and utilizing data integration tools.
- **Model Interpretability:** Ensuring that maintenance teams could interpret and trust the predictions required efforts in explaining model outputs. Transparent communication and training sessions were conducted to address this challenge.

Case Study 2: Demand Response Optimization in a Smart Grid
Background

A utility company implemented a demand response optimization system in its smart grid infrastructure to efficiently manage energy demand during peak periods.

Implementation

- **Smart Meter Data Analysis:** Leveraged big data analytics to analyze data from smart meters, considering historical usage patterns and external factors like weather forecasts.
- **Machine Learning for Demand Forecasting:** Developed machine learning models to forecast energy demand at various times of the day.
- **Dynamic Pricing Strategies:** Implemented dynamic pricing strategies based on real-time demand forecasts to incentivize users to shift their energy consumption.

Results

- **Reduced Peak Demand:** The demand response optimization system successfully reduced peak energy demand during high-usage periods.
- **Increased Grid Stability:** By distributing energy consumption more evenly, the smart grid experienced improved stability and reduced strain on the infrastructure.
- **User Engagement:** Users benefited from cost savings through dynamic pricing, leading to increased engagement with the demand response program.

Key Takeaways

- **User Incentives Matter:** Designing dynamic pricing models that provide clear incentives for users to shift their energy consumption can lead to successful demand response programs.
- **Real-time Data is Critical:** Timely and accurate data from smart meters is essential for effective demand forecasting and dynamic pricing.

Challenges and Solutions

- **User Education:** Overcoming user resistance to behavior change required a comprehensive education campaign. Clear communication about the benefits of demand response and how to participate was key.
- **Data Security and Privacy:** Addressing concerns about data security and privacy involved implementing robust encryption measures and transparent data usage policies.

Case Study 3: Grid Anomaly Detection in a Renewable Energy Network
Background

A utility operating a renewable energy network integrated big data analytics and machine learning to detect anomalies and ensure the stability of the grid in the presence of variable renewable energy sources.

Implementation

- **Data from Renewable Sources:** Collected data from various renewable sources, including solar panels and wind turbines.
- **Predictive Analytics for Energy Production:** Used big data analytics to predict energy production based on weather forecasts, historical data, and real-time conditions.
- **Anomaly Detection Models:** Trained machine learning models to detect anomalies in energy production or grid behavior.
- **Automated Response Mechanisms:** Integrated the anomaly detection system with automated response mechanisms to address identified issues.

Results

- **Enhanced Grid Stability:** Anomaly detection allowed for quick identification and resolution of issues, contributing to overall grid stability.
- **Optimized Renewable Energy Integration:** By predicting energy production more accurately, the utility optimized the integration of renewable energy into the grid.

Key Takeaways

- **Renewable Energy Integration:** Big data analytics and machine learning play a crucial role in managing the variability of renewable energy sources, ensuring a smooth transition to a more sustainable energy mix.
- **Automation for Rapid Response:** Automated response mechanisms, triggered by anomaly detection, enable utilities to respond rapidly to changing grid conditions.

Challenges and Solutions

- **Data Volume and Velocity:** Handling the high volume and velocity of data generated by renewable sources required scalable infrastructure and real-time processing capabilities.
- **Algorithm Robustness:** Ensuring the robustness of anomaly detection models involved continuous refinement based on feedback and ongoing monitoring.

FUTURE TRENDS AND EMERGING TECHNOLOGIES

The upcoming trends in big data analytics, machine learning, and mobile services for smart utilities are discussed in Table **7**.

Table 7. Trends.

SL. No.	Trends	Remarks
1	Enhanced Predictive Analytics	• Advancements in predictive analytics will lead to more accurate forecasting of energy demand, equipment failures, and grid behavior. • Improved models will enhance decision-making and resource allocation in smart utilities.
2	Explainable AI (XAI)	• Increased focus on making machine learning models more interpretable and transparent. • Explainable AI will be crucial for gaining user trust, addressing regulatory concerns, and ensuring the ethical use of AI in smart utility applications.
3	Edge AI for Real-time Processing	• The adoption of edge computing combined with AI capabilities (Edge AI) will enable real-time data processing at the edge of the network. • This trend will reduce latency, enhance responsiveness, and support time-sensitive applications in smart utility systems.
4	Autonomous Systems and Robotics	• Integration of autonomous systems and robotics powered by AI for tasks such as routine maintenance, inspections, and repairs in utility operations. • This trend will enhance operational efficiency and reduce manual intervention.
5	Experiential Mobile Services	• Evolution of mobile services towards more immersive and experiential interfaces. Augmented reality (AR) and virtual reality (VR) applications may be integrated into mobile services to enhance user engagement and provide interactive experiences.

The integration of big data analytics and machine learning in the context of secure and flexible mobile services for smart utilities in the 5G era is a cutting-edge and transformative approach. This convergence can lead to more efficient and intelligent management of utilities such as electricity, water, and gas. Here is an overview of how these technologies can be applied.

Data Collection and Integration

- **Smart Meters and Sensors:** Implementing smart meters and sensors to collect real-time data from utility infrastructures.
- **IoT Devices:** Utilizing Internet of Things (IoT) devices to gather data from various sources in the utility network.

Big Data Analytics

- **Data Processing:** Processing the massive volume of data generated by smart meters, sensors, and IoT devices.
- **Predictive Analytics:** Using historical data to predict consumption patterns, identify anomalies, and optimize resource allocation.

Machine Learning

- **Predictive Maintenance:** Employing machine learning algorithms to predict and prevent equipment failures, reducing downtime and maintenance costs.
- **Demand Forecasting:** Forecasting utility demand to optimize the distribution and allocation of resources.
- **Anomaly Detection:** Detecting unusual patterns or behaviors in the network that may indicate security threats or system malfunctions.

Secure Communication

- **Encryption and Authentication:** Ensuring secure communication channels between devices and the central system through robust encryption and authentication mechanisms.
- **Blockchain Technology:** Implementing blockchain for secure and transparent transactions and data sharing among stakeholders.

Flexibility in Service Provision

- **Dynamic Resource Allocation:** Adjusting the allocation of resources based on real-time demand and supply.
- **Adaptive Services:** Offering adaptive services that respond to changes in consumer behavior, environmental conditions, or regulatory requirements.

Integration with 5G Networks

- **Low Latency:** Leveraging the low-latency capabilities of 5G networks for real-time data processing and decision-making.
- **High Bandwidth:** Utilizing the high bandwidth of 5G to handle the massive data generated by smart utilities.

Security Measures

- **Cybersecurity Protocols:** Implementing robust cybersecurity protocols to protect against data breaches and cyber threats.
- **Regular Audits and Updates:** Conducting regular security audits and updates to ensure the system is resilient to evolving cyber threats.

The integration of big data analytics and machine learning with 5G technology in the context of smart utilities can significantly enhance the efficiency, security, and flexibility of mobile services in this domain. It allows for more intelligent decision-making, predictive maintenance, and a proactive approach to managing resources, ultimately contributing to a more sustainable and resilient utility infrastructure.

CONCLUSION

Reflecting on the significance of these technological advancements, it becomes evident that big data analytics, machine learning, and secure mobile services are instrumental in addressing the complex challenges faced by modern utility management. The adoption of 5G networks further amplifies the potential for real-time data processing and communication, ushering in an era of unprecedented connectivity and responsiveness. The significance lies not only in the operational efficiency and resource optimization brought about by these technologies but also in the ethical and responsible implementation that safeguards user privacy and societal well-being. The convergence of technology, regulation, and ethical considerations signifies a holistic and sustainable approach to smart utilities, creating a foundation for intelligent, adaptive, and secure services that cater to the evolving needs of communities. As smart utilities continue to evolve, striking a balance between technological innovation, regulatory compliance, and ethical considerations will be paramount. The journey toward smarter, more efficient, and secure utilities is not just a technological endeavor but a societal transformation, and the insights discussed in this article pave the way for a future where utility services are not only intelligent but also ethically and socially responsible.

REFERENCES

[1]   A. Z. Amiri, A. Heidari, N. J. Navimipour, M. Unal, and A. Mousavi, "Adventures in data analysis: A systematic review of Deep Learning techniques for pattern recognition in cyber-physical-social systems", *Multimedia Tools Appl.,* vol. 82, no. 1, pp. 1-65, 2023.

[2]   V.K. Mololoth, S. Saguna, and C. Åhlund, "Blockchain and machine learning for future smart grids: A review", *Energies,* vol. 16, no. 1, p. 528, 2023.
      [http://dx.doi.org/10.3390/en16010528]

[3]   S. Sahana, D. Singh, and I. Nath, "Importance of AI and ML towards smart sensor network utility enhancement", *Encyclopedia of Data Science and Machine Learning,* pp. 240-262, 2023.

[4]   İ. Yazici, I. Shayea, and J. Din, "A survey of applications of artificial intelligence and machine learning in future mobile networks-enabled systems", *Engineering Science and Technology, an International Journal,* vol. 44, p. 101455, 2023.

[5]   H. Xie *et al.*, "IntelliSense technology in the new power systems", *Renewable and Sustainable Energy Reviews,* vol. 177, p. 113229, 2023.

[6]   M.S. Abdalzaher, M.M. Fouda, H.A. Elsayed, and M.M. Salim, "Toward Secured IoT-Based Smart Systems Using Machine Learning", *IEEE Access,* vol. 11, pp. 20827-20841, 2023.
[http://dx.doi.org/10.1109/ACCESS.2023.3250235]

[7]   K. Thenmozhi, M. Pyingkodi, and K. Kanimozhi, "Effective load balance in iotsg with various machine learning techniques", *Smart Grids and Internet of Things: An Energy Perspective,* pp. 193-203, 2023.

[8]   A. Ullah *et al.*, "Smart cities: the role of Internet of Things and machine learning in realizing a data-centric smart environment", *Complex & Intelligent Systems,* vol. 9, no. 1, pp. 1-31, 2023.

[9]   Z. Ali *et al.*, "A generic internet of things (iot) middleware for smart city applications", *Sustainability,* vol. 15, no. 1, p. 743, 2023.

[10]   H. F. Ahmad *et al.*, "Leveraging 6G, extended reality, and IoT big data analytics for healthcare: A review", *Computer Science Review,* vol. 48, p. 100558, 2023.

[11]   K. Chakrapani, T. Kavitha, M. I. Safa, M. Kempanna, and B. Chakrapani, "Applications of artificial intelligence in intelligent combustion and energy storage technologies", *Applications of Big Data and Artificial Intelligence in Smart Energy Systems.* River Publishers, pp. 27-45, 2023.

[12]   M. Ryalat, H. ElMoaqet, and M. AlFaouri, "Design of a smart factory based on cyber-physical systems and Internet of Things towards Industry 4.0", *Appl. Sci. (Basel),* vol. 13, no. 4, p. 2156, 2023.
[http://dx.doi.org/10.3390/app13042156]

[13]   Z. Qadir, K.N. Le, N. Saeed, and H.S. Munawar, "Towards 6G Internet of Things: Recent advances, use cases, and open challenges", *ICT Express,* vol. 9, no. 3, pp. 296-312, 2023.
[http://dx.doi.org/10.1016/j.icte.2022.06.006]

[14]   T. Mazhar *et al.*, "Analysis of cyber security attacks and its solutions for the smart grid using machine learning and blockchain methods", *Future Internet,* vol. 15, no. 2, p. 83, 2023.

[15]   B. Ahmed *et al.*, "IoT based smart systems using artificial intelligence and machine learning: accessible and intelligent solutions", *2023 6th International Conference on Information Systems and Computer Networks (ISCON),* pp. 1-6, 2023.

[16]   S. Aminizadeh *et al.*, "The applications of machine learning techniques in medical data processing based on distributed computing and the Internet of Things", *Computer Methods and Programs in Biomedicine,* vol. 231, p. 107745, 2023.

[17]   A. Mishra *et al.*, "Emerging technologies and design aspects of next generation cyber physical system with a smart city application perspective", *International Journal of System Assurance Engineering and Management,* vol. 14, no. 3, pp. 699-721, 2023.

[18]   D.K. Nguyen, G. Sermpinis, and C. Stasinakis, "Big data, artificial intelligence and machine learning: A transformative symbiosis in favour of financial technology", *Eur. Financ. Manag.,* vol. 29, no. 2, pp. 517-548, 2023.
[http://dx.doi.org/10.1111/eufm.12365]

[19]   B. U. Maheswari *et al.*, "Internet of things and machine learning-integrated smart robotics", *Global Perspectives on Robotics and Autonomous Systems: Development and Applications,* pp. 240-258, 2023.

[20]   A. Rahman *et al.*, "Towards a blockchain-SDN-based secure architecture for cloud computing in smart industrial IoT", *Digital Communications and Networks,* vol. 9, no. 2, pp. 411-421, 2023.

[21]   M. M. H. Sifat *et al.*, "Towards electric digital twin grid: technology and framework review", *Energy and AI,* vol. 11, p. 100213, 2023.

[22]   H. Liao, E. Michalenko, and S.C. Vegunta, "Review of big data analytics for smart electrical energy systems", *Energies,* vol. 16, no. 8, p. 3581, 2023.
[http://dx.doi.org/10.3390/en16083581]

[23]   J. Leng *et al.*, "Towards resilience in Industry 5.0: A decentralized autonomous manufacturing paradigm", *Journal of Manufacturing Systems,* vol. 71, pp. 95-114, 2023.

[24]   I. A. T. Hashem *et al.*, "Urban Computing for Sustainable Smart Cities: Recent Advances, Taxonomy, and Open Research Challenges", *Sustainability,* vol. 15, no. 5, p. 3916, 2023.

[25]   F.M.M. Cirianni, A. Comi, and A. Quattrone, "Mobility Control Centre and Artificial Intelligence for Sustainable Urban Districts", *Information (Basel),* vol. 14, no. 10, p. 581, 2023.
[http://dx.doi.org/10.3390/info14100581]

[26]   J. Gu *et al.*, "Multistage quality control in manufacturing process using blockchain with machine learning technique", *Information Processing & Management,* vol. 60, no. 4, p. 103341.

[27]   *M. Syamala et al., "Machine Learning-Integrated IoT-Based Smart Home Energy Management System," in Handbook of Research on Deep Learning Techniques for Cloud-Based Industrial IoT.* IGI Global, 2023, pp. 219-235. [27]

[28]   H. Attou *et al.*, "Towards an intelligent intrusion detection system to detect malicious activities in cloud computing", *Applied Sciences,* vol. 13, no. 17, p. 9588, 2023.

[29]   M. A. Naeem *et al.*, "Cache in fog computing design, concepts, contributions, and security issues in machine learning prospective", *Digital Communications and Networks,* vol. 9, no. 5, pp. 1033-1052, 2023.

[30]   M.A. Naeem, Y.B. Zikria, R. Ali, U. Tariq, Y. Meng, and A.K. Bashir, "Cache in fog computing design, concepts, contributions, and security issues in machine learning prospective", *Digit. Commun. Netw.,* vol. 9, no. 5, pp. 1033-1052, 2023.
[http://dx.doi.org/10.1016/j.dcan.2022.08.004]

CHAPTER 8

An Overview of Computational Intelligence and Big Data Analytics for Smart Healthcare

Devasis Pradhan[1,*], **Tarique Akhtar**[2] and **Amit Kumar Sahoo**[3]

[1] *Department of Electronics & Communication Engineering, Acharya Institute of Technology, Bangalore, Karnataka, India*

[2] *Data Science Agility, Dubai*

[3] *Lead 1 Workforce Management, UST Global, Bangalore, India*

Abstract: Smart healthcare, propelled by technological advancements, is witnessing a paradigm shift in the way healthcare services are delivered. This paper explores the transformative impact of Computational Intelligence (CI) and Big Data Analytics on smart healthcare systems. Computational Intelligence encompasses artificial neural networks, fuzzy logic, genetic algorithms, and expert systems, while Big Data Analytics involves the processing and analysis of large datasets to extract meaningful insights. This integration aims to enhance the efficiency, accuracy, and personalized nature of healthcare delivery. The application of CI in smart healthcare includes disease diagnosis through medical image analysis and predictive analytics for identifying high-risk patients. Moreover, CI facilitates personalized medicine by tailoring treatment plans based on individual characteristics. On the other hand, Big Data Analytics contributes to clinical decision support, population health management, and real-time monitoring of patients. The combination of CI and Big Data Analytics enables the development of predictive models, decision support systems, and efficient utilization of data from Internet of Things (IoT) devices and sensors. However, the adoption of these technologies in smart healthcare is not without challenges. Privacy and security concerns surrounding patient data, interoperability issues, and ethical considerations demand careful attention. Establishing standards for data interoperability and addressing ethical concerns related to consent and algorithmic biases are imperative for the successful implementation of CI and Big Data Analytics in healthcare.

Keywords: Analytics, Bigdata, Clinical strategies, Computational intelligence, Healthcare.

INTRODUCTION

The healthcare landscape is undergoing a revolutionary transformation driven by technological advancements that promise to redefine patient care, optimize resou-

---

* **Corresponding author Devasis Pradhan:** Department of Electronics & Communication Engineering, Acharya Institute of Technology, Bangalore, Karnataka, India; E-mail: devasispradhan@acharya.ac.in

Devasis Pradhan, Mangesh M. Ghonge, Nitin S. Goje, Alessandro Bruno and Rajeswari (Eds.)

rce utilization, and enhance overall healthcare outcomes. At the forefront of this evolution are Computational Intelligence (CI) and Big Data Analytics, two pillars of innovation that synergistically contribute to the concept of smart healthcare. Smart healthcare leverages cutting-edge technologies to create a connected and intelligent ecosystem, enabling healthcare providers to deliver more personalized, efficient, and effective services. Computational Intelligence (CI) encompasses a suite of advanced algorithms inspired by human intelligence, including artificial neural networks, fuzzy logic, genetic algorithms, and expert systems. These computational techniques empower healthcare systems to analyze complex datasets, identify patterns, and make informed decisions, mirroring the cognitive processes of human professionals. Big Data Analytics, on the other hand, involves the systematic extraction of meaningful insights from vast and diverse datasets. In the context of healthcare, this includes clinical data, patient records, and real-time monitoring information. By harnessing the power of Big Data Analytics, healthcare practitioners can uncover valuable patterns, trends, and correlations that were previously obscured by the sheer volume and complexity of the data.

The integration of CI and Big Data Analytics in the healthcare domain holds the promise of addressing longstanding challenges and unlocking new possibilities. From early and accurate disease diagnosis to personalized treatment plans, and from real-time patient monitoring to predictive modeling, these technologies are reshaping the way healthcare is conceptualized, delivered, and experienced. This paper provides an in-depth overview of how Computational Intelligence and Big Data Analytics are being harnessed in the realm of smart healthcare. We explore their individual contributions, collaborative applications, and the transformative impact on patient care and healthcare management. Additionally, we delve into the challenges and considerations that must be addressed to ensure the ethical and secure implementation of these technologies in the complex and sensitive healthcare ecosystem. As we embark on this journey through the landscape of smart healthcare, it becomes evident that the fusion of computational intelligence and big data analytics is not merely a technological evolution; it is a pivotal revolution with the potential to redefine the future of healthcare delivery.

LITERATURE SURVEY

The adoption of Fourth Industrial Revolution technologies, particularly artificial intelligence, and big data has presented significant challenges for industries across various sectors [1]. In the healthcare industry, the adoption of these technologies has brought both immense benefits and numerous challenges. One of the major benefits is the ability to utilize artificial intelligence and big data analytics for smart healthcare. These technologies enable the analysis of vast amounts of healthcare data, including electronic health records, medical imaging, and genomics data, to extract valuable insights and support decision-making in healthcare settings. Furthermore, the combination of artificial intelligence and big data has the potential to revolutionize clinical practice by enabling predictive modeling, personalized medicine, and real-time monitoring of patients [2].

This advancement in technology allows for more accurate diagnoses, early detection of diseases, and the development of personalized treatment plans. The use of computational intelligence and big data analytics in smart healthcare has the potential to improve patient outcomes, enhance efficiency in healthcare delivery, and reduce healthcare costs. Additionally, the integration of these technologies allows for remote monitoring and telemedicine, expanding access to healthcare services for underserved populations. In summary, the integration of artificial intelligence and big data analytics in smart healthcare is expected to bring significant innovations by improving accuracy in diagnosis, enabling predictive modeling for patient outcomes, and enhancing the overall efficiency of healthcare delivery. Furthermore, the use of computational intelligence and big data analytics in smart healthcare can also contribute to the early detection of epidemics and disease outbreaks, as well as improve population health management by identifying trends and patterns in large-scale health data.

Overall, the adoption of artificial intelligence and big data analytics in smart healthcare has the potential to revolutionize the industry by improving patient care, expanding access to healthcare services, and driving cost savings. The integration of artificial intelligence and big data analytics in smart healthcare has the potential to revolutionize the industry by improving patient outcomes, enhancing efficiency in healthcare delivery, and reducing costs [3]. By leveraging the power of computational intelligence and big data analytics, healthcare providers can make more accurate diagnoses, personalize treatment plans, and improve patient outcomes. Furthermore, the real-time monitoring and analysis of patient data can lead to early detection of diseases and timely interventions. This can ultimately result in improved population health management and overall healthcare quality [4].

ROLE OF COMPUTATIONAL INTELLIGENCE IN SMART HEALTHCARE

Computational intelligence plays a significant role in the development and implementation of smart healthcare. It involves the use of various techniques and algorithms, such as machine learning, deep learning, and natural language processing, to analyze large volumes of data from different sources [5 - 7]. These intelligent systems can learn from the data and make predictions, recommendations, and decisions to support healthcare providers in their decision-making process. One of the key areas where computational intelligence has been applied in smart healthcare is in the analysis of electronic health records (EHRs). EHRs contain a wealth of patient data, including medical history, diagnoses, medications, and lab results. By applying machine learning algorithms to EHRs, healthcare providers can identify patterns and trends in patient data that can help in diagnosing diseases, predicting outcomes, and developing treatment plans [8, 9]. This not only improves the accuracy and speed of diagnosis but also reduces the risk of errors due to human bias.

Another crucial aspect of smart healthcare is remote patient monitoring. With the rise of wearables and other internet-connected devices, patients can now track their health data, such as heart rate, blood pressure, and glucose levels, in real-time [10]. Computational intelligence is used to analyze this data and identify any abnormal patterns or trends that may require medical attention. This allows for early detection and intervention, reducing the risk of complications and hospital readmissions. Moreover, computational intelligence is also being used in predictive analytics in smart healthcare. Fig. (**1**) discussed about various component of CI. By analyzing large sets of patient data, such as demographics, lifestyle, and medical history, machine learning algorithms can predict a patient's risk of developing certain diseases. This information can be used to develop personalized preventive measures and interventions, such as lifestyle changes and screenings, to mitigate the risk of developing these diseases [11 - 15].

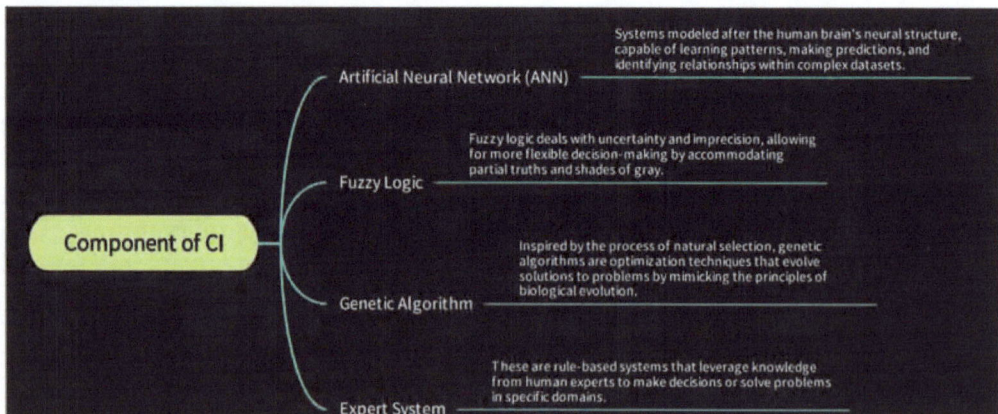

Fig. (1). Various components of Computational Intelligent (CI).

IMPACT OF COMPUTATIONAL INTELLIGENCE IN SMART HEALTHCARE

The use of computational intelligence in smart healthcare has had a significant impact on the healthcare industry. One of the key benefits is the improvement in the quality of patient care. By analyzing large amounts of data, intelligent systems can identify patterns and trends that may not be apparent to healthcare providers, leading to more accurate diagnoses and treatment plans. Fig. (**2**) shows about the impact of CI. This results in better health outcomes for patients. Moreover, the use of computational intelligence in smart healthcare has also improved the efficiency of healthcare services [16, 17]. By automating tasks such as data analysis and decision-making, healthcare providers can save time and resources, allowing them to focus on providing better care to patients. This also reduces the risk of errors and improves the speed of diagnosis and treatment, leading to better patient satisfaction. Another significant impact of computational intelligence in smart healthcare is the reduction of healthcare costs. By identifying health issues at an early stage and preventing complications, the need for expensive treatments and hospitalizations can be reduced. Additionally, the use of remote patient monitoring and predictive analytics can help in identifying high-risk patients and providing them with preventive care, reducing overall healthcare costs [18 - 20].

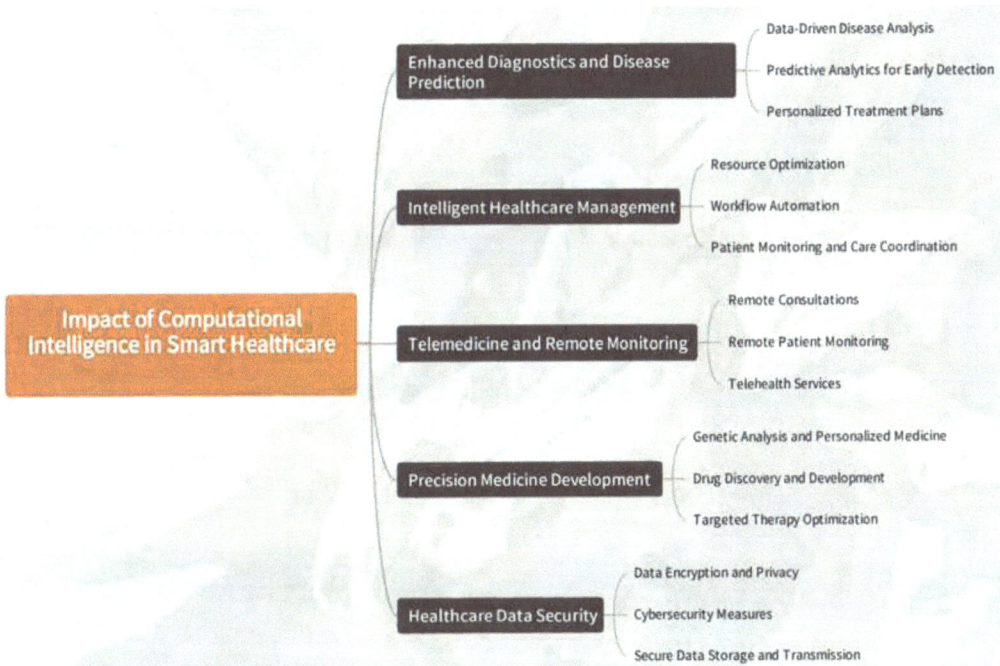

Fig. (2). Impact of CI in smart healthcare.

BIG DATA ANALYTICS IN SMART HEALTHCARE

Big Data Analytics refers to the process of examining, processing, and interpreting vast and diverse datasets to extract valuable insights, patterns, and trends. In the context of healthcare, it involves the use of advanced analytics techniques to derive meaningful information from large volumes of clinical, administrative, and patient-generated data. Fig. (**2**). shows various components of BDA in the Health care sector. Fig. (**3**) shows various component of BDA.

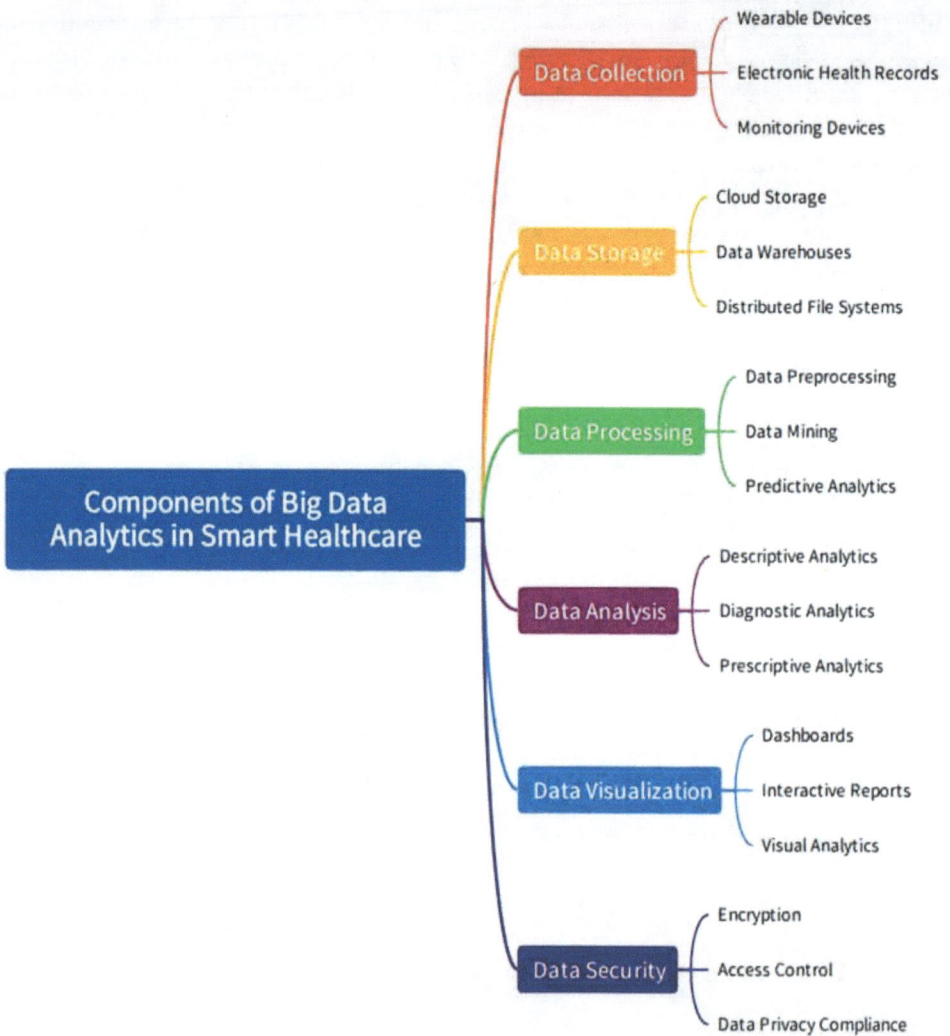

Fig. (3). Components of BDA in smart health care.

The use of big data analytics in smart healthcare has several key benefits. First and foremost, it allows healthcare providers to have a more comprehensive view of a patient's health. By analyzing data from various sources, healthcare professionals can better understand a patient's medical history, lifestyle, and potential risk factors. This information can help them make more accurate diagnoses and develop personalized treatment plans [21 - 23]. For example, a patient's electronic health records can be combined with data from their fitness tracker to provide a complete picture of their health and identify any potential health issues early on. In addition to improving patient care, big data analytics can also help healthcare providers streamline their operations and reduce costs. By analyzing data on patient outcomes and treatment effectiveness, providers can identify areas for improvement and implement more efficient practices. This can lead to reduced hospital readmissions, shorter hospital stays, and overall cost savings for both patients and healthcare organizations. For instance, big data analytics can be used to predict which patients are at risk of readmission, allowing healthcare providers to intervene and prevent unnecessary hospital visits. Another major benefit of big data analytics in smart healthcare is its ability to assist in disease prevention and control. By analyzing data from various sources, healthcare providers can identify patterns and trends in disease outbreaks and quickly respond to potential public health crises. For example, big data analytics can help track the spread of infectious diseases, such as the recent COVID-19 pandemic, and aid in identifying high-risk areas and populations. This information can then be used to develop targeted interventions and prevent further spread of the disease. Table 1 shows the various applications of BDA in smart healthcare [24, 25].

IOT DEVICES AND SENSORS IN SMART HEALTHCARE

IoT devices and sensors are used in various aspects of healthcare, from patient monitoring to supply chain management. These devices and sensors are embedded in different medical equipment, wearables, and even in hospital infrastructure, enabling the collection and exchange of data [26]. For instance, wearables such as fitness trackers and smartwatches can track the vital signs, activity levels, and sleep patterns of patients. Medical equipment like blood pressure monitors, glucometers, and ECG machines can also be connected to the internet, allowing for real-time data collection. In addition to patient monitoring, IoT devices and sensors are also utilized in healthcare facilities to improve operational efficiency [27]. For example, sensors can be placed in hospital rooms to monitor temperature, humidity, and air quality, ensuring a comfortable and safe environment for patients. IoT devices can also be used to track the location and usage of medical equipment, reducing the time spent searching for items and preventing stock shortages. The vast amount of data collected by IoT devices and sensors in healthcare is known as big data. This data is often unstructured, coming from various sources and in different formats. Big data analytics is the process of extracting meaningful insights from this data and leveraging it to improve decision-making and drive innovation [28]. In smart healthcare, big data analytics plays a crucial role in transforming raw data into actionable insights. Fig. (4) depicts IoT Sensors in smart healthcare.

Table 1. Applications of BDA in smart healthcare.

Sl. No.	Area	Specific Work	Remarks
1	Clinical Decision Support	Patient Risk Stratification	Big Data Analytics aids in identifying high-risk patients by analyzing historical and real-time data. This facilitates proactive interventions and personalized treatment plans.
		Evidence-Based Medicine	Clinical decision support systems leverage big data to provide healthcare professionals with evidence-based insights, improving the accuracy and efficacy of medical decisions.
2	Population Health Management	Disease Surveillance	Big Data Analytics allows for the monitoring and tracking of disease patterns across populations, enabling healthcare providers to implement targeted interventions and allocate resources efficiently.
		Predictive Modeling	By analyzing large datasets, healthcare organizations can create predictive models to forecast disease outbreaks, assess population health risks, and plan preventive measures.
	Real-time Patient Monitoring	Remote Patient Monitoring	Big Data Analytics facilitates real-time monitoring of patients through wearable devices and sensors. This enables healthcare professionals to track vital signs, medication adherence, and overall patient well-being outside traditional healthcare settings.
		Early Warning Systems	Analyzing streaming data from various sources helps in the early detection of deteriorating patient conditions, allowing for timely interventions and hospital readmission prevention.

Role of IoT Devices and Sensors in Big Data Analytics

The integration of IoT devices and sensors in healthcare has significantly expanded the scope and quality of data available for analysis. These devices generate real-time data, providing a continuous stream of information that can be used for predictive analytics. For instance, wearable devices can monitor a patient's heart rate, blood oxygen levels, and other vital signs, providing real-time data that can be used to detect anomalies and predict potential health issues [29]. Moreover, IoT devices and sensors can collect data from multiple sources, providing a holistic view of a patient's health. This data can be combined with other relevant information, such as electronic health records, to gain a deeper understanding of a patient's health status and history. Fig. (**5**) shows role of IoT devices in BDA. This comprehensive data can then be analyzed using big data analytics tools and techniques to identify patterns, trends, and correlations that can help healthcare professionals make more informed decisions.

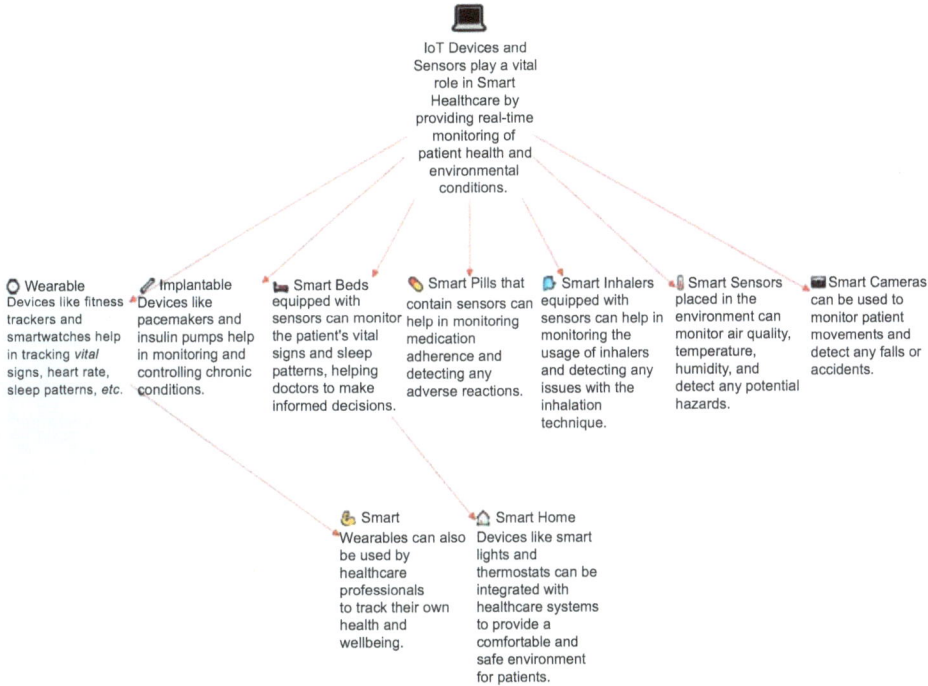

IoT Devices and Sensors play a vital role in Smart Healthcare by providing real-time monitoring of patient health and environmental conditions.

○ Wearable Devices like fitness trackers and smartwatches help in tracking *vital* signs, heart rate, sleep patterns, *etc.*

Implantable Devices like pacemakers and insulin pumps help in monitoring and controlling chronic conditions.

Smart Beds equipped with sensors can monitor the patient's vital signs and sleep patterns, helping doctors to make informed decisions.

Smart Pills that contain sensors can help in monitoring medication adherence and detecting any adverse reactions.

Smart Inhalers equipped with sensors can help in monitoring the usage of inhalers and detecting any issues with the inhalation technique.

Smart Sensors placed in the environment can monitor air quality, temperature, humidity, and detect any potential hazards.

Smart Cameras can be used to monitor patient movements and detect any falls or accidents.

Smart Wearables can also be used by healthcare professionals to track their own health and wellbeing.

Smart Home Devices like smart lights and thermostats can be integrated with healthcare systems to provide a comfortable and safe environment for patients.

Fig. (4). IoT devices and sensors in smart healthcare.

Benefits of IoT Devices and Sensors in Smart Healthcare

The use of IoT devices and sensors in conjunction with big data analytics offers several benefits in smart healthcare. These include:

- **Improved Patient Care:** The continuous data collection and analysis enabled by IoT devices and sensors can provide healthcare professionals with real-time insights into a patient's health status, allowing for timely interventions and personalized treatment plans.
- **Cost Reduction:** By tracking the location and usage of medical equipment and optimizing supply chain management, IoT devices, and sensors can help reduce operational costs for healthcare facilities.
- **Enhanced Efficiency:** IoT devices and sensors can automate various tasks, such as collecting patient data and monitoring equipment, freeing healthcare professionals to focus on more critical tasks.

- **Early Detection and Prevention of Diseases:** The combination of IoT devices, big data analytics, and artificial intelligence can help identify patterns and correlations in patient data, enabling early detection and prevention of diseases.

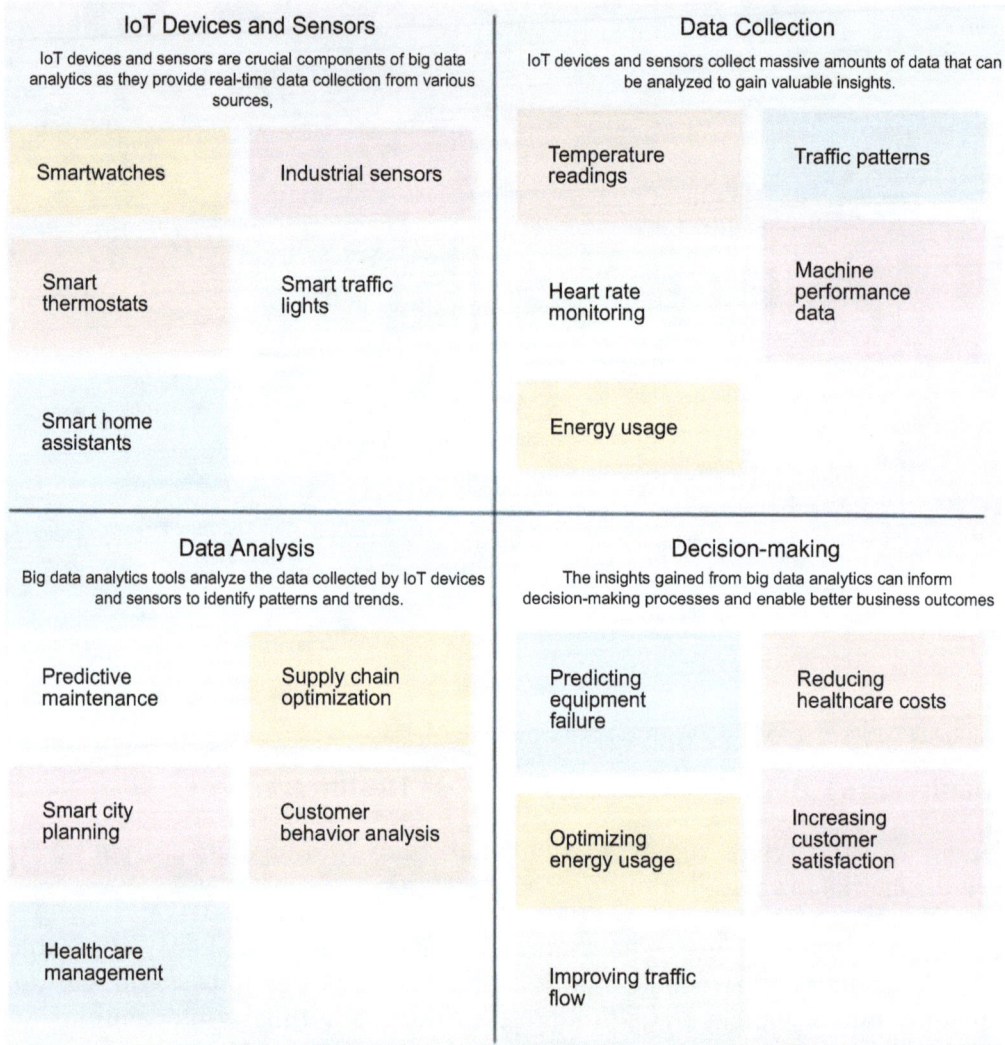

IoT Devices and Sensors

IoT devices and sensors are crucial components of big data analytics as they provide real-time data collection from various sources,

Smartwatches	Industrial sensors
Smart thermostats	Smart traffic lights
Smart home assistants	

Data Collection

IoT devices and sensors collect massive amounts of data that can be analyzed to gain valuable insights.

Temperature readings	Traffic patterns
Heart rate monitoring	Machine performance data
Energy usage	

Data Analysis

Big data analytics tools analyze the data collected by IoT devices and sensors to identify patterns and trends.

Predictive maintenance	Supply chain optimization
Smart city planning	Customer behavior analysis
Healthcare management	

Decision-making

The insights gained from big data analytics can inform decision-making processes and enable better business outcomes

Predicting equipment failure	Reducing healthcare costs
Optimizing energy usage	Increasing customer satisfaction
Improving traffic flow	

Fig. (5). Role of IoT devices and sensors in BDA.

INTEGRATION OF CI AND BIG DATA IN SMART HEALTHCARE

The integration of computational intelligence and big data in smart healthcare has revolutionized the healthcare industry and has the potential to improve patient outcomes, reduce costs, and enhance the overall quality of care [30].

Computational intelligence, which is a subfield of artificial intelligence, involves the use of algorithms and mathematical models to analyze and interpret complex data, while big data refers to the vast amount of data that is generated and collected by various sources such as electronic health records, medical devices, and wearables. The healthcare industry has been generating massive amounts of data for decades, but it is only in recent years that the technology has advanced enough to effectively analyze and utilize this data. This is where the integration of computational intelligence and big data comes into play [31] (Fig. **6**). By combining these two powerful technologies, healthcare providers can gain valuable insights into patient health, make more accurate diagnoses, and develop personalized treatment plans.

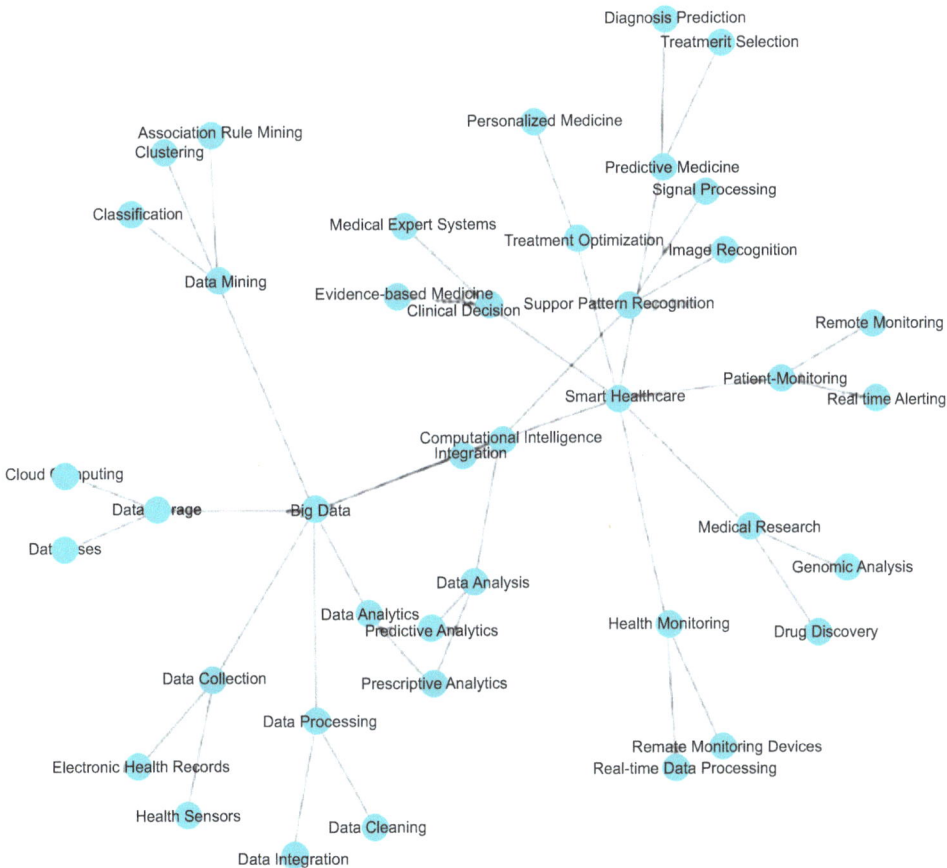

Fig. (6). Various integrating factors.

One of the major benefits of this integration is the ability to make predictions and identify patterns in patient data. With the help of computational intelligence algorithms, healthcare providers can analyze large datasets to identify trends and risk factors for diseases. This can lead to early detection and prevention of diseases, ultimately saving lives. For example, by analyzing a patient's electronic health records and genetic data, computational intelligence algorithms can predict the likelihood of developing diseases such as diabetes or heart disease. Another area where this integration is making a significant impact is in disease diagnosis [32, 33]. With the help of big data and computational intelligence, healthcare providers can now make more accurate and timely diagnoses. By analyzing a patient's symptoms, medical history, and genetic data, algorithms can provide a list of possible diagnoses, helping doctors to make a more informed decision. This can be especially helpful in cases where a patient's symptoms may not fit into a traditional diagnosis, or when a patient has a rare disease. In addition to improving diagnosis and prediction, the integration of computational intelligence and big data is also enhancing the treatment process. With the help of big data analytics, healthcare providers can develop personalized treatment plans for each patient based on their unique health data. This can lead to more effective and efficient treatment, reducing the risk of adverse reactions and improving patient outcomes. Moreover, this integration is also improving the management of chronic diseases. By continuously monitoring and analyzing patient data, healthcare providers can detect any changes in a patient's condition and intervene before it becomes a serious issue. This can lead to a reduction in hospital readmissions and better management of chronic conditions, resulting in improved overall health outcomes and reduced healthcare costs [34]. The use of computational intelligence and big data in smart healthcare is not limited to patient care. It is also being utilized in healthcare operations to improve efficiency and reduce costs. For example, predictive analytics can be used to forecast patient demand, allowing healthcare facilities to allocate resources accordingly. This can help reduce wait times, improve patient satisfaction, and optimize resource utilization. However, the integration of computational intelligence and big data in smart healthcare also raises concerns about patient privacy and data security. With the collection and analysis of large amounts of sensitive patient data, there is a risk of data breaches and misuse. Therefore, it is crucial for healthcare providers to have strict data privacy and security measures in place to protect patient information. The integration of computational intelligence and big data in smart healthcare has the potential to transform the healthcare industry. By providing valuable insights and improving patient care, this integration is helping healthcare providers make more informed decisions and deliver better outcomes for patients [35].

CHALLENGES AND CONSIDERATION

In today's rapidly advancing healthcare industry, Continuous Intelligence and Big Data Analytics have emerged as powerful tools for improving patient outcomes and optimizing healthcare processes. However, the implementation of CI and Big Data Analytics in smart healthcare comes with its own set of challenges that need to be addressed effectively. One of the major challenges in implementing CI and Big Data Analytics in smart healthcare is data security and privacy concerns. With the vast amount of sensitive patient data being collected and analyzed, ensuring the security and privacy of this data is crucial. This challenge can be addressed by implementing robust data security measures, such as encryption techniques, access controls, and secure data storage systems. Interoperability is another challenge in implementing CI and Big Data Analytics in smart healthcare. Healthcare systems and devices often use different standards and formats for data storage and sharing, making it difficult to integrate and analyze data from multiple sources. To address this challenge, establishing interoperability standards and implementing data integration strategies can facilitate seamless data exchange and analysis across different healthcare systems and devices. Ethical considerations are also important when implementing CI and Big Data Analytics in smart healthcare. The use of patient data for analysis and decision-making should adhere to ethical principles, such as informed consent, privacy protection, and transparency. Table **2** discusses the challenges and specific considerations.

Table 2. Challenges and consideration.

Sl.No.	Issues	Challenges	Consideration
1	Data Security and Privacy Concerns	• Healthcare data is sensitive and subject to privacy regulations, making it a prime target for cyber threats. • Ensuring the security of patient information and maintaining compliance with data protection laws pose significant challenges.	• Implement robust encryption techniques to safeguard data during transmission and storage. • Adhere strictly to regulatory frameworks such as HIPAA (Health Insurance Portability and Accountability Act) and GDPR (General Data Protection Regulation). • Conduct regular security audits and vulnerability assessments to identify and address potential weaknesses.

(Table 2) cont.....

2	Interoperability Issues	• Healthcare systems often use diverse technologies and standards, leading to interoperability challenges when integrating CI and Big Data Analytics solutions. • Incompatible data formats and systems hinder seamless collaboration.	• Adopt standardized data formats and interoperability protocols to facilitate communication between different systems. • Promote the use of open standards in healthcare information exchange. • Encourage collaboration and communication between stakeholders to align technological infrastructures.
3	Ethical Considerations	• The use of CI and Big Data in healthcare raises ethical concerns related to consent, transparency, bias, and the potential misuse of sensitive information.	• Establish clear and transparent consent mechanisms, ensuring patients are informed about how their data will be used. • Implement ethical guidelines for data collection, analysis, and storage. • Regularly assess algorithms for biases and take corrective measures to mitigate any disparities.
4	Data Quality and Accuracy	• Inaccurate or incomplete data can compromise the effectiveness of CI and Big Data Analytics applications, leading to erroneous conclusions and decisions.	• Implement data quality assurance measures, including validation and cleansing processes. • Educate healthcare professionals on the importance of accurate data entry and maintenance. • Continuously monitor and update datasets to ensure relevance and reliability.
5	Resource Allocation and Cost	• The implementation of CI and Big Data Analytics requires significant financial investments in technology, training, and infrastructure.	• Conduct a thorough cost-benefit analysis to justify investments and demonstrate the potential return on investment. • Seek partnerships and collaborations to share resources and reduce costs. • Prioritize phased implementations to manage budget constraints effectively.
6	Lack of Skilled Professionals	• The demand for skilled professionals in CI and Big Data Analytics often outpaces the supply, leading to a shortage of qualified personnel.	• Invest in training programs for existing staff to enhance their skills. • Foster partnerships with educational institutions to promote the development of a skilled workforce. • Utilize external consultants and experts to supplement in-house expertise.

CONCLUSION

The field of healthcare has seen significant advancements in recent years, with the emergence of computational intelligence and big data analytics playing a crucial role. These technologies have the potential to revolutionize the way healthcare is delivered, managed, and experienced. In this paper, we have explored the key findings and insights from the overview of these technologies and their transformative potential in shaping the future of smart healthcare. We have also discussed the ethical and technical considerations necessary for their successful and responsible implementation in the healthcare domain. One of the key findings from the overview is the increasing amount of data being generated in the healthcare sector. The use of electronic health records, wearable devices, and other digital tools has resulted in the generation of large volumes of data. This data can be harnessed through big data analytics to gain valuable insights and improve healthcare outcomes. Moreover, the integration of computational intelligence with big data analytics has led to the development of advanced algorithms and predictive models, enabling healthcare professionals to make more accurate diagnoses and treatment plans. Another crucial insight from the overview is the potential of these technologies to improve the efficiency and cost-effectiveness of healthcare delivery. By analyzing large datasets, healthcare organizations can identify patterns and trends, which can help them make informed decisions about resource allocation, supply chain management, and other operational processes. This can lead to significant cost savings and improved productivity. Additionally, the use of remote monitoring and telehealth solutions, powered by computational intelligence and big data analytics, can reduce the need for in-person visits, thereby saving time and resources for both patients and healthcare providers. One of the most significant transformations that computational intelligence and big data analytics can bring to the healthcare sector is the shift from reactive to proactive care. With the help of predictive models, healthcare professionals can identify potential health risks and intervene early to prevent disease progression. This can lead to better health outcomes and reduced healthcare costs in the long run. Moreover, personalized medicine is becoming a reality with the use of these technologies. By combining patient data, from genetic information to lifestyle habits, with advanced algorithms, healthcare providers can tailor treatment plans to individual patients, leading to better treatment outcomes.

REFERENCES

[1]   G. Dicuonzo, F. Donofrio, A. Fusco, and M. Shini, "Healthcare system: Moving forward with artificial intelligence", *Technovation,* vol. 120, p. 102510, 2023.
[http://dx.doi.org/10.1016/j.technovation.2022.102510]

[2]   J. Ahn, J. Lee, S. Yoon, and J.K. Choi, "A novel resolution and power control scheme for energy-

efficient mobile augmented reality applications in mobile edge computing", *IEEE Wirel. Commun. Lett.,* vol. 9, no. 6, pp. 750-754, 2020.
[http://dx.doi.org/10.1109/LWC.2019.2950250]

[3]   A. Panicacci, "Does expressing emotions in the local language help migrants acculturate?", *Int. J. Lang. Cult.,* vol. 6, no. 2, pp. 279-304, 2019.
[http://dx.doi.org/10.1075/ijolc.17013.pan]

[4]   N. Renugadevi, S. Saravanan, and C.M. Naga Sudha, "Revolution of smart healthcare materials in big data analytics", *Mater. Today Proc.,* vol. 81, pp. 834-841, 2023.
[http://dx.doi.org/10.1016/j.matpr.2021.04.256]

[5]   R.F. Mansour, J. Escorcia-Gutierrez, M. Gamarra, V.G. Díaz, D. Gupta, and S. Kumar, "Artificial intelligence with big data analytics-based brain intracranial hemorrhage e-diagnosis using CT images", *Neural Comput. Appl.,* vol. 35, no. 22, pp. 16037-16049, 2023.
[http://dx.doi.org/10.1007/s00521-021-06240-y]

[6]   X. Chen, H. Xie, Z. Li, G. Cheng, M. Leng, and F.L. Wang, "Information fusion and artificial intelligence for smart healthcare: a bibliometric study", *Inf. Process. Manage.,* vol. 60, no. 1, p. 103113, 2023.
[http://dx.doi.org/10.1016/j.ipm.2022.103113]

[7]   S. AlZu'bi *et al.*, "Diabetes monitoring system in smart health cities based on big data intelligence", *Future Internet,* vol. 15, no. 2, p. 85, 2023.

[8]   P. Kumar, S. Chauhan, and L.K. Awasthi, "Artificial intelligence in healthcare: Review, ethics, trust challenges & future research directions", *Eng. Appl. Artif. Intell.,* vol. 120, p. 105894, 2023.
[http://dx.doi.org/10.1016/j.engappai.2023.105894]

[9]   S. Bag, P. Dhamija, R.K. Singh, M.S. Rahman, and V.R. Sreedharan, "Big data analytics and artificial intelligence technologies based collaborative platform empowering absorptive capacity in health care supply chain: An empirical study", *J. Bus. Res.,* vol. 154, p. 113315, 2023.
[http://dx.doi.org/10.1016/j.jbusres.2022.113315]

[10]   O. Ali *et al.*, "A systematic literature review of artificial intelligence in the healthcare sector: Benefits, challenges, methodologies, and functionalities", *Journal of Innovation & Knowledge,,* vol. 8, no. 1, p. 100333, 2023.

[11]   A.K. Sangaiah, S. Rezaei, A. Javadpour, and W. Zhang, "Explainable AI in big data intelligence of community detection for digitalization e-healthcare services", *Appl. Soft Comput.,* vol. 136, p. 110119, 2023.
[http://dx.doi.org/10.1016/j.asoc.2023.110119]

[12]   N. Jiwani, K. Gupta, and P. Whig, "Machine learning approaches for analysis in smart healthcare informatics", *Machine Learning and Artificial Intelligence in Healthcare Systems,* pp. 129-154, 2023.

[13]   T.A. Suleiman, and A. Adinoyi, "Telemedicine and smart healthcare—The role of artificial intelligence, 5G, cloud services, and other enabling technologies", *Int. J. Commun. Netw. Syst. Sci.,* vol. 16, no. 3, pp. 31-51, 2023.
[http://dx.doi.org/10.4236/ijcns.2023.163003]

[14]   P. V. Thayyib *et al.*, "State-of-the-art of artificial intelligence and big data analytics reviews in five different domains: A bibliometric summary", *Sustainability,* vol. 15, no. 5, p. 4026.

[15]   S.A. Suha, and T.F. Sanam, "Exploring dominant factors for ensuring the sustainability of utilizing artificial intelligence in healthcare decision making: An emerging country context", *International Journal of Information Management Data Insights,* vol. 3, no. 1, p. 100170, 2023.
[http://dx.doi.org/10.1016/j.jjimei.2023.100170]

[16]   T. Huynh-The *et al.*, "Artificial intelligence for the metaverse: A survey", *Engineering Applications of Artificial Intelligence,* vol. 117, p. 105581, 2023.

[17]   S. Kumar, W.M. Lim, U. Sivarajah, and J. Kaur, "Artificial intelligence and blockchain integration in

business: Trends from a bibliometric-content analysis", *Inf. Syst. Front.*, vol. 25, no. 2, pp. 871-896, 2023.
[PMID: 35431617]

[18]   C. Guo and J. Chen, "Big data analytics in healthcare", In: *Knowledge Technology and Systems: Toward Establishing Knowledge Systems Science.* Springer Nature Singapore: Singapore, 2023, pp. 27-70.

[19]   A. Amjad, P. Kordel, and G. Fernandes, "A review on innovation in healthcare sector (telehealth) through artificial intelligence", *Sustainability (Basel),* vol. 15, no. 8, p. 6655, 2023.
[http://dx.doi.org/10.3390/su15086655]

[20]   S. Baker and W. Xiang, "Artificial Intelligence of Things for smarter healthcare: A survey of advancements, challenges, and opportunities", *IEEE Commun. Surv. Tutor.,* 2023.

[21]   N. Khalid, A. Qayyum, M. Bilal, A. Al-Fuqaha, and J. Qadir, "Privacy-preserving artificial intelligence in healthcare: Techniques and applications", *Comput. Biol. Med.,* p. 106848, 2023.

[22]   V. Chamola *et al.*, "Artificial intelligence-assisted blockchain-based framework for smart and secure EMR management", *Neural Computing and Applications,* vol. 35, no. 31, pp. 22959-22969, 2023.

[23]   B. Alhayani *et al.*, "5G standards for the Industry 4.0 enabled communication systems using artificial intelligence: Perspective of smart healthcare system", *Applied Nanoscience,* vol. 13, no. 3, pp. 1807-1817, 2023.

[24]   E. Morrow *et al.*, "Artificial intelligence technologies and compassion in healthcare: A systematic scoping review", *Frontiers in Psychology,* vol. 13, p. 971044, 2023.

[25]   Y. Zhu, Q. Yang, and X. Mao, "Global trends in the study of smart healthcare systems for the elderly: Artificial intelligence solutions", *International Journal of Computational Intelligence Systems,* vol. 16, no. 1, p. 105, 2023.
[http://dx.doi.org/10.1007/s44196-023-00283-w]

[26]   S. A. Alowais *et al.*, "Revolutionizing healthcare: The role of artificial intelligence in clinical practice", *BMC Medical Education,* vol. 23, no. 1, p. 689, 2023.

[27]   C.H. Lee, C. Wang, X. Fan, F. Li, and C.H. Chen, "Artificial intelligence-enabled digital transformation in elderly healthcare field: Scoping review", *Adv. Eng. Inform.,* vol. 55, p. 101874, 2023.
[http://dx.doi.org/10.1016/j.aei.2023.101874]

[28]   B. Khan *et al.*, "Drawbacks of artificial intelligence and their potential solutions in the healthcare sector", *Biomedical Materials and Devices,* pp. 1-8, 2023.

[29]   M. A. I. Mozumder *et al.*, "Metaverse for digital anti-aging healthcare: An overview of potential use cases based on artificial intelligence, blockchain, IoT technologies, its challenges, and future directions", *Applied Sciences,* vol. 13, no. 5, p. 5127, 2023.

[30]   M. M. Hasan *et al.*, "Review on the evaluation and development of artificial intelligence for COVID-19 containment", *Sensors,* vol. 23, no. 1, p. 527, 2023.

[31]   M. T. Ho *et al.*, "Understanding the acceptance of emotional artificial intelligence in Japanese healthcare system: A cross-sectional survey of clinic visitors' attitude", *Technology in Society,* vol. 72, p. 102166, 2023.

[32]   L. P. Vishwakarma, R. K. Singh, R. Mishra, and A. Kumari, "Application of artificial intelligence for resilient and sustainable healthcare system: Systematic literature review and future research directions", *Int. J. Prod. Res.,* pp. 1-23, 2023.

[33]   A. V. L. N. Sujith *et al.*, "Systematic review of smart health monitoring using deep learning and artificial intelligence", *Neuroscience Informatics,* vol. 2, no. 3, p. 100028, 2022.

[34]   P. Manickam *et al.*, "Artificial intelligence (AI) and Internet of Medical Things (IoMT) assisted biomedical systems for intelligent healthcare", *Biosensors,* vol. 12, no. 8, p. 562, 2022.

[35]   S. Manickam, S. A. Mariappan, S. M. Murugesan, S. Hansda, A. Kaushik, R. Shinde, and S. P. Thipperudraswamy, "Artificial intelligence (AI) and Internet of Medical Things (IoMT) assisted biomedical systems for intelligent healthcare", *Biosensors,* vol. 12, no. 8, p. 562, 2022.

CHAPTER 9

Identification and Interconnection of Symptoms of Hypertension using Interpretive Structural Model: A Qualitative Survey

Varsha Umesh Ghate[1,*], **Sachin Kadam**[2], **Umesh Ghate**[3], **Anupam Mukherjee**[4] and **Anita Sardar Patil**[1]

[1] *Bharati Vidyapeeth (Deemed to be University) Homoeopathic Medical College, Pune, India*

[2] *Institute of Management and Entrepreneurship Development, Bharati Vidyapeeth (Deemed to be University), Pune, India*

[3] *Bharati Vidyapeeth (Deemed to be University), College of Ayurved, Pune, India*

[4] *Department of Health and Family Welfare, West Bengal Homoeopathic Health Service, Government of West Bengal, India*

Abstract: Hypertension (HTN) is one of the major global public health maladies. Equally, the impact on the incidence of hypertension in smart cities is increasing due to the abundant use of electromagnetic fields like 5G. HTN may not have any warning indications so the interconnection of its symptoms is crucial for early diagnosis and management. Thus, in order to examine a set of symptoms and how they relate to one another in HTN, the authors employed interpretive structural model (ISM). In the first stage, the authors identified a total of 18 symptoms of hypertension by review. After an interview with the expert panel, 17 additional symptoms were found in the second stage. In the third stage, expert panel members were asked to rate the symptoms with a score 1 to 4. The authors used an ISM in the fourth stage to develop a causality rule-base for the diagnosis of hypertension. Any combination of symptoms, such as 1. Dizziness followed by a) Chest pain + Palpitation + Transient chest pain after exertion /or, b) Headache + Fainting. 2. Headache followed by a) Chest pain + Palpitation + Transient chest pain after exertion /or, b) Dizziness + Fainting. 3. Fainting followed by a) Chest pain + Palpitation + Transient chest pain after exertion, /or, b) Dizziness + Headache, may be used to identify hypertension. It was discovered that the presence of nosebleed symptoms did not contribute to the hypertension diagnosis. Data analytics is a common tool used by smart cities to enhance healthcare facilities. By contributing insights into the early detection of hypertension throughout smart cities, the ISM model can support data-driven decision-making and enhance the healthcare system.

* **Corresponding author Varsha Umesh Ghate:** Bharati Vidyapeeth (Deemed to be University) Homoeopathic Medical College, Pune, India; E-mail: varshaghate29@gmail.com

Devasis Pradhan, Mangesh M. Ghonge, Nitin S. Goje, Alessandro Bruno and Rajeswari (Eds.)

Keywords: Hypertension symptoms, Hypertension management, Interpretive structural model (ISM), Qualitative survey study, Qualitative research methods, Symptom interconnection.

INTRODUCTION

Hypertension (HTN) is one of the silent killers, contributing to cardiovascular diseases and early death worldwide. Uncontrolled hypertension results in a number of prevalent complications, such as coronary heart disease, peripheral vascular disease, congestive heart failure, renal insufficiency, and stroke [1, 2]. Equally, the impact on the incidence of hypertension in smart cities is increasing due to the abundant use of electromagnetic fields like 5G. As hypertension may not have any warning indications, the majority of people with the condition are unaware of the issue. The early detection of hypertension and the interrelation of its symptoms are therefore crucial for the timely diagnosis and management of HTN [2]. ISM, which stands for Interpretative Structural Modeling, the ISM technique, Warfield first developed in 1974, was created to handle complex challenges [3]. It allows the development of a map of intricate connections within several factors of convoluted conditions by either an individual or group of experts, that can be utilized to gain fresh perspectives and devise renewed solutions to the present problem [4, 5]. To create a visual representation of the scenario, ISM uses concept synthesis, transitive logic, and pair-wise comparison [6]. Many famous organizations, including NASA, have utilized this approach to resolve challenging problems [7]. Thus, in order to examine a set of symptoms and how they relate to one another in hypertension, the authors employed interpretive structural modelling (Fig. **1**).

First Stage: Screening and Identification of HTN Symptoms

To identify HTN symptoms, a literature review of modern literature, medical textbooks, and databases including Scopus, Science Direct, Web of Science, PubMed, Research Gate, Google Scholar, and Shodhganga was conducted. And so, after doing an extensive review, the authors identified a total of 18 symptoms, including blood in your urine (hematuria), blurred or unclear vision, headache, dizziness, fatigue, nosebleed, chest pain, shortness of breath, mental confusion, fainting, heart palpitations, nausea and/or vomiting, tinnitus (ringing in ears), sleepiness, insomnia, confusion, excess sweating, and sometimes patients are asymptomatic [8 - 19].

Second Stage: Confirmation and Verification of HTN Symptoms

Symptoms identified in the first stage by the extensive review were again confirmed and verified by an interview with ten expert panel members having eligibility for a minimum of i) Ten years of teaching/ research experience at the college/university/ industries OR ii) Ten years of experience as a practitioner. Thus, the authors from the expert panel members identified a total of seventeen new symptoms, including projectile vomiting, giddiness, tiredness, weakness in limbs, convulsions, unconsciousness, transient chest pain after exertion, chest pain relief after rest, oliguria, pain at epigastric region, generalized anasarca, swelling on face, swelling on eyelids, swelling on hand, swelling on abdominal wall, swelling on legs and swelling of vulva.

METHOD:

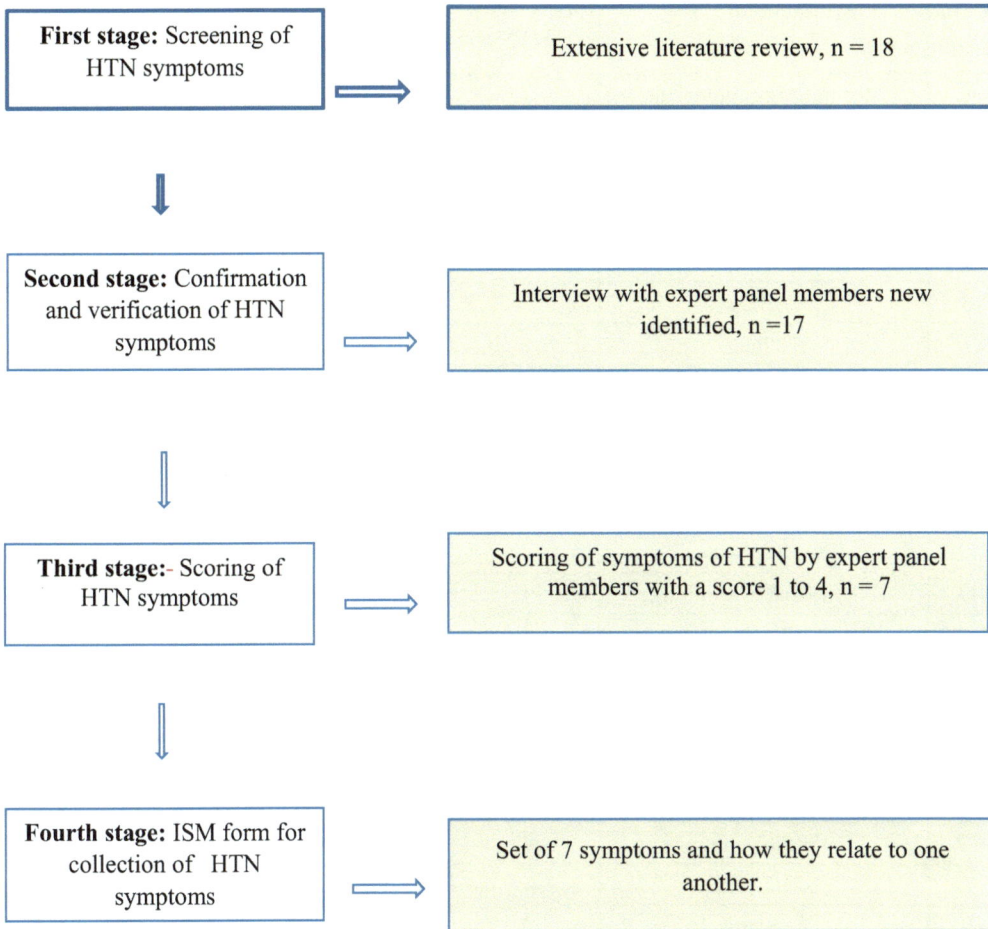

First stage: Screening of HTN symptoms ⟹ Extensive literature review, n = 18

Second stage: Confirmation and verification of HTN symptoms ⟹ Interview with expert panel members new identified, n = 17

Third stage:- Scoring of HTN symptoms ⟹ Scoring of symptoms of HTN by expert panel members with a score 1 to 4, n = 7

Fourth stage: ISM form for collection of HTN symptoms ⟹ Set of 7 symptoms and how they relate to one another.

Fig. (1). The study was carried out by the use of four-stage technique.

Third Stage: Scoring of HTN Symptoms

In the third stage, 10 expert panel members were interviewed once more and asked to rate each symptom using the following scale 1 = Not relevant to HTN, 2 = Somewhat relevant to HTN, 3 = Quite relevant to HTN, and 4 = Highly relevant to HTN (Table **1**).

Table 1. Scoring of symptoms of HTN by expert panel members [2, 3].

Sr. No.	Symptoms of HTN	1= Not Relevant to HTN	2= Somewhat Relevant to HTN	3= Quite Relevant to HTN	4= Highly Relevant to HTN
1	Most times asymptomatic	1	3	5	1
2	Blood in your urine (hematuria)	4	4	1	1
3	Blurry vision or double vision	-	4	5	1
4	Chest pain	-	1	7	2
5	Shortness of breath	1	4	4	1
6	Dizziness	-	-	7	3
7	Fatigue	-	5	3	2
8	Mental fog	2	2	5	1
9	Nose bleed	-	1	4	5
10	Headache	-	-	1	9
11	Fainting	-	1	4	5
12	Heart palpitations	-	1	4	5
13	Nausea and/or vomiting	1	2	4	3
14	Tinnitus (Ringing in Ears)	1	5	2	2
15	Sleepiness	3	3	3	1
16	Insomnia	3	4	3	-
17	Confusion	1	5	2	2
18	Excess sweating	1	3	2	4
19	Projectile vomiting	1	4	2	3
20	Giddiness	-	3	4	3
21	Tiredness	2	1	5	2
22	Weakness in limbs	1	2	3	4
23	Convulsions	3	2	1	4
24	Unconsciousness	4	-	3	3
25	Transient chest pain after exertion	1	1	2	6

(Table 1) cont.....

Sr. No.	Symptoms of HTN	1= Not Relevant to HTN	2= Somewhat Relevant to HTN	3= Quite Relevant to HTN	4= Highly Relevant to HTN
26	Chest pain relives after rest	3	3	3	1
27	Oliguria	1	5	4	-
28	Pain at epigastric region	3	4	3	-
29	Generalized anasarca	3	1	5	1
30	Swelling on face	4	2	2	2
31	Swelling on eyelids	1	3	3	3
32	Swelling on hand	2	3	4	1
33	Swelling on abdominal wall	5	4	1	-
34	Swelling on legs	1	3	2	4
35	Swelling of vulva	7	2	1	-

Fourth Stage: ISM form for Collection of HTN Symptoms

In this study in order to examine a set of symptoms and how they relate to one another in hypertension, the authors employed ISM (Table **2**). Consequently, a qualitative ISM approach is thought to be appropriate for the objective of this study as available literature is not sufficiently far-reaching to early diagnosis of hypertension by a set of symptoms and the link between them. In order to collect interview data, the ISM form has been developed. Twenty experts on a panel were interviewed by the authors.

Table 2. Evaluation set of symptoms and how they relate to one another after Interview [4, 5].

Sr. No.	Symptoms	Chest Pain	Dizziness	Nose Bleed	Headache	Fainting	Palpitation	Transient Chest Pain After Exertion
1	Chest pain	-	0	-	0	0	1	1
2	Dizziness	0	-	-	1	1	0	0
3	Nose bleed	-	0	-	-	0	-	-
4	Headache	1	1	-	-	0	0	0
5	Fainting	0	1	-	0	-	0	0
6	Palpitation	1	0	-	0	0	-	1
7	Transient chest pain after exertion	1	0	-	0	0	1	-

= Blocked areas.

DATA ANALYSIS

Experts

Total number of experts interviewed = 20, Threshold to consider representative qualitative opinion = 50% (>= 10 experts), Threshold table by representing value 1 for above threshold and value 0 for below threshold.

Structural Self-Interaction Matrix (SSIM)

Symbols to define relationships in SSIM: V → row variable influences corresponding column variable, A→ row variable is influenced by corresponding column variable, X → row and corresponding column variable influence each other, O → row and corresponding column variable have no relationship.

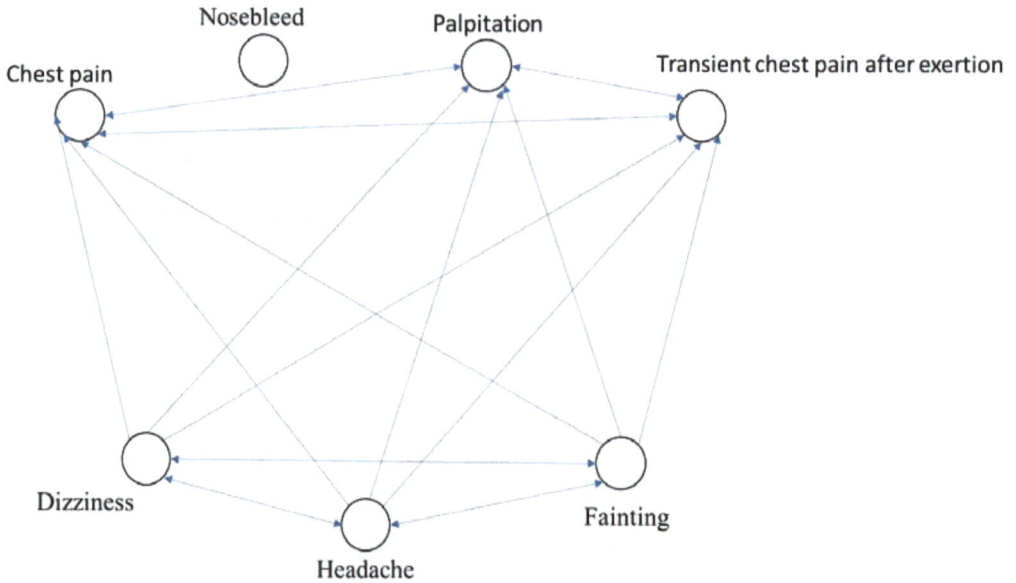

Fig. (2). ISM model.

RESULTS

Interpretation of ISM model to formulate causality rule-base towards the diagnosis of hypertension condition. Condition of hypertension may be diagnosed if a person displays any one of the primary symptoms followed by any one of the corresponding secondary symptoms. Any combination of symptoms, such as 1. Dizziness followed by a) Chest pain + Palpitation + Transient chest pain after exertion /or, b) Headache + Fainting. 2. Headache followed by a) Chest pain + Palpitation + Transient chest pain after exertion /or, b) Dizziness + Fainting. 3.

Fainting followed by a) Chest pain + Palpitation + Transient chest pain after exertion, /or, b) Dizziness + Headache, may be used to identify hypertension. It was discovered that the presence of nosebleed symptoms did not contribute to the hypertension diagnosis.

DISCUSSION

Information and communications technology (ICT), which includes wireless communication for mobile phones and, for example, Wi-Fi employing electromagnetic fields (EMF), has developed in smart cities at a rate never seen in previous decades. In addition, the widespread use of electromagnetic fields like 5G is having an increasing impact on the prevalence of hypertension in smart cities [20]. The incidence of HTN-related issues in the Indian healthcare system is significantly impacted by late diagnosis of the condition, as many patients are unaware that they have HTN. Uncontrolled hypertension poses a serious health risk. Beliefs regarding the signs and symptoms of hypertension can negatively impact treatment compliance [21]. Therefore, early detection of hypertension and an understanding of how its symptoms relate to one another are essential for prompt diagnosis and treatment of HTN.

The technique known as interpretive structural modeling (ISM) turns vague and poorly expressed thoughts of systems into observable, precisely defined models that may be applied to a variety of instances. ISM aids in deciphering the logic behind the connections between system components and their order. ISM provides a framework for clinicians to organize information, and to help them make the best decisions for treatment [21].

So this research tried to implement the same to rule out a set of symptoms and how they relate to one another in hypertension, the authors employed interpretive structural modeling. A total number of 20 experts were interviewed by authors. The threshold to consider representative qualitative opinion = 50% (>= 10 experts). Symbols to define relationships in the Structural Self-Interaction Matrix (SSIM) were V → row variable influences corresponding column variable, A → row variable is influenced by corresponding column variable, X → row and corresponding column variable influence each other, O → row and corresponding column variable have no relationship (Table 3). By applying the ISM model for HTN, we obtained a rule-based set of symptoms by which early diagnosis of HTN is possible for physicians (Table 5). The ISM model further discovered that the presence of nosebleed symptoms did not contribute to the HTN diagnosis. By implementing ISM model in HTN symptoms, researchers can increase the awareness about early diagnosis of hypertension by physicians to avoid further HTN complications.

Table 3. Structural self-interaction matrix (SSIM) [6].

Variables 1 2 3 4 5 6 7
Chest pain O O A O X X
Dizziness O X X O O
Nosebleed O O O O
Headache O O O
Fainting O O
Palpitation X
Transient chest pain after exertion

Table 4. Reduced conical matrix (CM) [7].

Variables	1	3	6	7	2	4	5	Driving Power	Level
Chest pain	1	0	1	1	0	0	0	3	1
Nosebleed	0	1	0	0	0	0	0	1	1
Palpitation	1	0	1	1	0	0	0	3	1
Transient chest pain after exertion	1	0	1	1	0	0	0	3	1
Dizziness	1*	0	1*	1*	1	1	1	6	2
Headache	1	0	1*	1*	1	1	1*	6	2
Fainting	1*	0	1*	1*	1	1*	1	6	2
Dependence Power	6	1	6	6	3	3	3	-	-
Level	1	1	1	1	2	2	2	-	-

Table 5. Set of symptoms for early diagnosis of HTN [8].

Sr. No.	Primary Symptom	Secondary Symptom
1	Dizziness	1. Chest pain + Palpitation + Transient chest pain after exertion 2. Headache + Fainting
2	Headache	1. Chest pain + Palpitation + Transient chest pain after exertion 2. Dizziness + Fainting
3	Fainting	1. Chest pain + Palpitation + Transient chest pain after exertion 2. Dizziness + Headache

Note: Nosebleed symptom found to be irrelevant with respect to the diagnosis of hypertension condition.

CONCLUSION

The identification and interconnection of symptoms of hypertension using an Interpretive Structural Model (ISM) can be a valuable approach to better understand the complex relationships between different symptoms and early diagnosis of hypertension for its management. By contributing insights into the early detection of hypertension throughout smart cities, the ISM model can support data-driven decision-making and enhance the healthcare system.

REFERENCES

[1]   C.J.L. Murray, and A.D. Lopez, "Mortality by cause for eight regions of the world: Global Burden of Disease Study", *Lancet,* vol. 349, no. 9061, pp. 1269-1276, 1997.
[http://dx.doi.org/10.1016/S0140-6736(96)07493-4] [PMID: 9142060]

[2]   "The sixth report of the Joint National Committee on prevention, detection, evaluation, and treatment of high blood pressure", *Archives of internal medicine,* vol. 157, no. 21, pp. 2413-2446, 1997.
[http://dx.doi.org/10.1001/archinte.157.21.2413]

[3]   J.N. Warfield, "Toward interpretation of complex structural models", *IEEE Trans. Syst. Man Cybern.,* vol. SMC-4, no. 5, pp. 405-417, 1974.
[http://dx.doi.org/10.1109/TSMC.1974.4309336]

[4]   T.W. Sheu, T.L. Chen, C.P. Tsai, J.W. Tzeng, C.P. Deng, and M. Nagai, "Analysis of students' misconception based on rough set theory", *Journal of Intelligent Learning Systems and Applications,* vol. 5, no. 2, pp. 67-83, 2013.
[http://dx.doi.org/10.4236/jilsa.2013.52008]

[5]   P. S. Poduval, and V. R. Pramod, "Interpretive Structural Modeling (ISM) and its application in analyzing factors inhibiting implementation of Total Productive Maintenance (TPM)", *Int. J. Qual. Reliab. Manage.,* vol. 32, no. 3, pp. 308-331, 2015.
[http://dx.doi.org/10.1108/IJQRM-06-2013-0090]

[6]   M. Nilashi, M. Dalvi, O. Ibrahim, M. Zamani, and T. Ramayah, "An interpretive structural modelling of the features influencing researchers' selection of reference management software", *J. Librarian. Inform. Sci.,* vol. 51, no. 1, pp. 34-46, 2019.
[http://dx.doi.org/10.1177/0961000616668961]

[7]   A. Digalwar, R.D. Raut, V.S. Yadav, B. Narkhede, B.B. Gardas, and A. Gotmare, "Evaluation of critical constructs for measurement of sustainable supply chain practices in lean-agile firms of Indian origin: A hybrid ISM-ANP approach", *Bus. Strategy Environ.,* vol. 29, no. 3, pp. 1575-1596, 2020.
[http://dx.doi.org/10.1002/bse.2455]

[8]   S. Singh, R. Shankar, and G.P. Singh, "Prevalence and associated risk factors of hypertension: A cross-sectional study in urban varanasi", *Int. J. Hypertens.,* vol. 2017, pp. 1-10, 2017.
[http://dx.doi.org/10.1155/2017/5491838] [PMID: 29348933]

[9]   C.J. Bulpitt, C.T. Dollery, and S. Carne, "Change in symptoms of hypertensive patients after referral to hospital clinic", *Heart,* vol. 38, no. 2, pp. 121-128, 1976.
[http://dx.doi.org/10.1136/hrt.38.2.121] [PMID: 1259826]

[10]   K.I. Kjellgren, J. Ahlner, B. DahlÖF, H. Gill, T. Hedner, and R. SÄljÖ, "Perceived symptoms amongst hypertensive patients in routine clinical practice – a population-based study", *J. Intern. Med.,* vol. 244, no. 4, pp. 325-332, 1998.
[http://dx.doi.org/10.1046/j.1365-2796.1998.00377.x] [PMID: 9797496]

[11]   M. Reiff, S. Schwartz, and M. Northridge, "Relationship of depressive symptoms to hypertension in a household survey in Harlem", *Psychosom. Med.,* vol. 63, no. 5, pp. 711-721, 2001.

[http://dx.doi.org/10.1097/00006842-200109000-00002] [PMID: 11573017]

[12] A.K. Goodhart, "Hypertension from the patient's perspective", *Br. J. Gen. Pract.,* vol. 66, no. 652, p. 570, 2016.
[http://dx.doi.org/10.3399/bjgp16X687757] [PMID: 27789496]

[13] R. Chen, K. Dharmarajan, V.T. Kulkarni, N. Punnanithinont, A. Gupta, B. Bikdeli, P.S. Mody, and I. Ranasinghe, "Most important outcomes research papers on hypertension", *Circ. Cardiovasc. Qual. Outcomes,* vol. 6, no. 4, pp. e26-e35, 2013.
[http://dx.doi.org/10.1161/CIRCOUTCOMES.113.000424] [PMID: 23838106]

[14] A.M. Iqbal, and S.F. Jamal, "Essential Hypertension", In: *StatPearls.* StatPearls Publishing, 2023.

[15] P. K. Whelton, R. M. Carey, W. S. Aronow, D. E. Casey Jr, K. J. Collins, C. Dennison Himmelfarb, S. M. DePalma, S. Gidding, K. A. Jamerson, D. W. Jones, E. J. MacLaughlin, P. Muntner, B. Ovbiagele, S. C. Smith Jr, C. C. Spencer, R. S. Stafford, S. J. Taler, R. J. Thomas, K. A. Williams Sr, J. D. Williamson, and J. T. Wright, ACC/AHA/AAPA/ABC/ACPM/AGS/APhA/ASH/ASPC/NMA/PCNA, "Guideline for the Prevention, Detection, Evaluation, and Management of High Blood Pressure in Adults: Executive Summary: A Report of the American College of Cardiology/American Heart Association Task Force on Clinical Practice Guidelines", *Hypertension (Dallas, Tex.: 1979),* vol. 71, no. 6, pp. 1269-1324, 2018.
[http://dx.doi.org/10.1161/HYP.0000000000000066]

[16] D.M. Rabi, K.A. McBrien, R. Sapir-Pichhadze, M. Nakhla, S.B. Ahmed, S.M. Dumanski, S. Butalia, A.A. Leung, K.C. Harris, L. Cloutier, K.B. Zarnke, M. Ruzicka, S. Hiremath, R.D. Feldman, S.W. Tobe, T.S. Campbell, S.L. Bacon, K.A. Nerenberg, G.K. Dresser, A. Fournier, E. Burgess, P. Lindsay, S.W. Rabkin, A.P.H. Prebtani, S. Grover, G. Honos, J.E. Alfonsi, J. Arcand, F. Audibert, G. Benoit, J. Bittman, P. Bolli, A.M. Côté, J. Dionne, A. Don-Wauchope, C. Edwards, T. Firoz, J.Y. Gabor, R.E. Gilbert, J.C. Grégoire, S.E. Gryn, M. Gupta, F. Hannah-Shmouni, R.A. Hegele, R.J. Herman, M.D. Hill, J.G. Howlett, G.L. Hundemer, C. Jones, J. Kaczorowski, N.A. Khan, L.M. Kuyper, M. Lamarre-Cliche, K.L. Lavoie, L.A. Leiter, R. Lewanczuk, A.G. Logan, L.A. Magee, B.K. Mangat, P.A. McFarlane, D. McLean, A. Michaud, A. Milot, G.W. Moe, S.B. Penner, A. Pipe, A.Y. Poppe, E. Rey, M. Roerecke, E.L. Schiffrin, P. Selby, M. Sharma, A. Shoamanesh, P. Sivapalan, R.R. Townsend, K. Tran, L. Trudeau, R.T. Tsuyuki, M. Vallée, V. Woo, A.D. Bell, and S.S. Daskalopoulou, "Hypertension canada's 2020 comprehensive guidelines for the prevention, diagnosis, risk assessment, and treatment of hypertension in adults and children", *Can. J. Cardiol.,* vol. 36, no. 5, pp. 596-624, 2020.
[http://dx.doi.org/10.1016/j.cjca.2020.02.086] [PMID: 32389335]

[17] M. Middeke, B. Lemmer, B. Schaaf, and L. Eckes, "Prevalence of hypertension-attributed symptoms in routine clinical practice: a general practitioners-based study", *J. Hum. Hypertens.,* vol. 22, no. 4, pp. 252-258, 2008.
[http://dx.doi.org/10.1038/sj.jhh.1002305] [PMID: 18007681]

[18] Williams, Bryan, "Essential hypertension—definition, epidemiology, and pathophysiology", in David A. Warrell, Timothy M. Cox, and John D. Firth (eds), Oxford Textbook of Medicine, 5 ed, Oxford Textbooks (Oxford, 2010; online edn, Oxford Academic), 29 Oct. 2015.
[http://dx.doi.org/10.1093/med/9780199204854.003.161701_update_001]

[19] T. Unger, C. Borghi, F. Charchar, N. A. Khan, N. R. Poulter, D. Prabhakaran, A. Ramirez, M. Schlaich, G. S. Stergiou, M. Tomaszewski, R. D. Wainford, B. Williams, and A. E. Schutte, "International Society of Hypertension Global Hypertension Practice Guidelines", *Hypertension (Dallas, Tex),* vol. 75, no. 6, pp. 1334-1357, 2020.
[http://dx.doi.org/10.1161/HYPERTENSIONAHA.120.15026]

[20] F. Amiri, M. Moradinazar, J. Moludi, Y. Pasdar, F. Najafi, E. Shakiba, B. Hamzeh, and A. Saber, "The association between self-reported mobile phone usage with blood pressure and heart rate: evidence from a cross-sectional study", *BMC Public Health,* vol. 22, no. 1, p. 2031, 2022.
[http://dx.doi.org/10.1186/s12889-022-14458-1] [PMID: 36344963]

[21] G. Granados-Gámez, J.G. Roales-Nieto, A. Gil-Luciano, E. Moreno-San Pedro, and V.V. Márquez-Hernández, "A longitudinal study of symptoms beliefs in hypertension", *Int. J. Clin. Health Psychol.,* vol. 15, no. 3, pp. 200-207, 2015.
[http://dx.doi.org/10.1016/j.ijchp.2015.07.001] [PMID: 30487837]

CHAPTER 10

Health Terminology Standards: A Comparative Study for the Patient Complaint Translation System

Bhanudas Suresh Panchbhai[1,*] and **Varsha Makarand Pathak**[2]

[1] *Department of Computer Science, R.C. Patel Arts, Commerce and Science College, Shirpur, Maharashtra, India*

[2] *Department of Computer Applications, KCES'S Institute of Management and Research, Maharashtra, India*

Abstract: When providing an acceptable diagnosis, health terminology helps with the right usage of language to describe illnesses, ailments, and symptoms in patients. There could be serious repercussions for the healthcare industry if this specification is unclear. Uniformity in medical terminology becomes essential when discussing the integration of automation and artificial intelligence into the scenario. The International Standards Organization (ISO) states that terminologies should be formal groupings of concepts that are linguistically unconnected, with a preferred name, suitable synonyms, and links between the concepts clearly expressed for each concept. To assist decision support systems, data sharing between health information systems, epidemiological analysis, research to support health services research, administrative task management, and other activities, standard terminology should be utilized in an electronic health record (EHR). This study examines ten popular clinical terminologies, including LOINC, NDC, and SNOMED Clinical Terms (SNOMED- CT), along with their histories, purposes, kinds, and structures. They also consist of CDT, CPT, RxNorm, HCPCS-Level II, and ICD-10, as well as ICD-10-CM, ICD-10-PCS, and ICD-10. Each criterion's advantages and disadvantages will be considered in this investigation. A comparative analysis is conducted by analyzing multiple terminologies to identify the advantages and disadvantages of each one separately.

Keywords: CPT, CDT, Data transmission, Electronic health records, HCPCS-Level II, ICD-10, LOINC, NDU, RxNORM, SNOMED-CT, Terminology.

---

* **Corresponding author Bhanudas Suresh Panchbhai:** Department of Computer Science, R.C. Patel Arts, Commerce and Science College, Shirpur, Maharashtra, India; E-mail: Bharat.panchbhai@gmail.com

Devasis Pradhan, Mangesh M. Ghonge, Nitin S. Goje, Alessandro Bruno and Rajeswari (Eds.)

INTRODUCTION

The objective of this research is to analyze health terminologies and create a comparison table that summarizes each according to several factors, such as owner, purpose, applications, area of use, *etc.* Which one is more appropriate for clinical terminology is determined by that finding.

Due to the current COVID-19 situation, a pandemic has been declared by the World Health Organization (WHO). In this pandemic situation, the healthcare industry is being revived in many aspects. The field is undergoing technology intervention with the realization of the importance of data in an appropriate form, systematic handling of data sources such as EHRs (Electronic Health Records), sensors, and other sources like medical ontology. Healthcare organizations want to digitize processes but do not unnecessarily disturb established clinical workflows. Therefore, we now have as much as 80 percent of data unstructured and of poor quality. This brings us to a pertinent challenge of data extraction and utilization in the healthcare space through natural language processing (NLP).

This data is as it is today and given the amount of time and effort it would need for humans to read and reformat it, thus, we cannot yet make effective decisions in healthcare through analytics because of the unstructured form of available data. Therefore, there is a higher need to leverage this unstructured data as we shift from a fee-for-service healthcare model to value-based care. NLP, or natural language processing, is useful in such a scenario. Automatic computational processing of human languages is referred to as "natural language processing" (NLP). This includes both algorithms that receive a real person-written text as input and algorithms that produce natural-looking text as outputs. NLP-based chatbots already possess the capabilities of well and truly mimicking human behavior and executing a myriad of tasks. When it comes to implementing the same on a much larger use case, like a hospital it can be used to parse information and extract critical strings of data, thereby offering an opportunity for us to leverage unstructured data (Alarsh Tiwari, 2020) [1]. Interoperability in information technology refers to linking many systems and effectively exchanging data between them. Every healthcare organization keeps its patients' medical records in accordance with either local or globally recognized standards. By preventing duplications and delays, these data and health message standards support their system's ability to communicate data and make it interoperable (Benson and Grieve, 2016) [2]. This study focuses on translating patients' chief complaints in the Marathi language into relevant medical terminology. For this, we need to analyze various standard health terminologies.

Clinical terminologies come in a variety of forms and are used in healthcare services. In this study, the most used ten clinical terminologies will be examined. The ICD-10 Procedure Coding System, Current Procedural Terminology (CPT), Logical Observation Identifiers Names and Codes (LOINC), Nomenclature for Properties and Units (NPU), and SNOMED Clinical Terms (SNOMED CT), ICD-10; RxNORM; NDU; and CDT are defined in the introduction and the remaining section of this paper is structured as follows: The background is given in Section II . Section III includes a table summarizing the ten most crucial health language standards, with associated terms. The discussion is presented in part IV, and a conclusion is provided in section V based on a study of the literature and a comparison of ten different clinical terminologies.

BACKGROUND

Both nationally and internationally, the necessity for standardized terminology in the healthcare industry has long been acknowledged. The International Statistical Institute first used the International Statistical Classification of Diseases (ICD) to classify causes of mortality in 1893 [3]. The ICD was expanded to cover morbidity in 1948, when the World Health Organization took over management of it following World War II [4]. In the same year that all countries agreed to adopt the 6th Decennial Revision of the International Lists of Diseases and Causes of Death, the suggestion to create national committees was made to assist in addressing "...statistical problems in the fields of health and vital statistics for study by national technicians as a preliminary step in the international development of standards and methods." [5]. To accomplish this in the United States, the National Committee on Vital and Health Statistics (NCVHS) was established in 1949. Numerous organizations, including the Institute of Medicine (IOM; now the National Academies of Sciences, Engineering, and Medicine), have studied the healthcare sector over the past 30 years and reported on the importance of adopting standardized terminologies and vocabularies for enhancing the delivery and quality of healthcare. The IOM revised and updated its 1991 recommendations regarding the computer-based patient record in 1997, noting that vocabulary standards were required to guarantee the integrity of clinical data in electronic health records (EHRs), as well as its retrieval, interpretation, and exchange, and that progress in improving health care would be difficult to achieve without an increase in the scope, use, and automation of the patient record [6].

Since the inception of Medicare and Medicaid, the use of standardized terminology for reporting and payment in healthcare settings in the United States has developed (enacted 1965). It took longer than anticipated to produce the nationwide standardized billing and claim forms used by the programmes, as well as the associated codes. For instance, the American Medical Association (AMA) released the Current Procedural Terminology (CPT) codes for physician services for the first time in 1966, but it took until 1978 for the use of CPT to be included in the larger Healthcare Common Procedural Coding System (HCPCS) [7]. HCPCS was developed in order to offer a uniform coding system for the description of covered services and goods.

Diagnosis Related Groups (DRG) was created as a new future payment system to serve as the foundation of Medicare's hospital reimbursement system as part of payment reform in the 1980s to address rising hospital expenses (Social Security Amendments of 1983) [8]. DRG is an ICD morbidity and procedure-based statistical classification system that pays a fixed sum based on the typical cost of care for hospital services. Given that DRGs rely on the clinically ambiguous ICD-9-CM, concerns have been expressed concerning this new system [9]. The Presidential Advisory Commission on Consumer Protection and Quality in the Health Care Industry was founded in 1996 as a result of rising healthcare expenses and growing worries about healthcare access and quality. In order to enable advancements in information systems, the commission suggested that national standards be established for the classification, categorization, and organization of health information.

The commission's other suggestions led to the creation of the National Quality Forum (NQF) in 1999. The NQF has an organized procedure for measuring endorsement . It was developed to bring consistency and agreement to the formulation of health quality measures used by both public and private sectors. Since 2012, NQF has taken an increasing interest in the discussion surrounding "eMeasures." Standard clinical vocabularies including SNOMED CT, LOINC, and RxNorm as well as ICD-10-CM are used by eMeasures, which are designed to be used in combination with electronic health records [10].

HEALTH TERMINOLOGY STANDARDS

For this investigation, the following definitions from the Patient Medical Record Information (PMRI) report of the National Committee on Vital and Health Statistics (NCVHS) are appropriate [11]:

Terminology is a continuum of code set, classification, and nomenclature (or vocabulary).

Code is a phrase given to make it simpler to understand. Generally speaking, a coding system is mentioned in most computer processing terminology. A code set is only a list of codes and the words and phrases to which they are assigned.

Grouping similar or related concepts together for easy retrieval is the process of classification. A few examples of how concepts could be placed in a categorization system are with respect to main categories, alphabetically, historically, or numerically. Nomenclature or Vocabulary A group of specialized phrases reduce or eliminate ambiguity to enable accurate communication.

"Controlled vocabulary" denotes simply the set of individual phrases in the vocabulary.

The following Fig. (1) offers a key framework for understanding the various levels of health IT from a technological perspective:

- Application Level Electronic Prescription (e-prescribing), Clinical Decision Support (CDS), Electronic Medication Administration Records (eMAR), Results Reporting, Electronic Documentation, Interface Engines, *etc.*
- Messaging Standards at the Communication Level HL7, ADT, NCPDP, X12, DICOM, ASTM, and others.
- Coding Standards like LOINC, ICD-9, CPT, NDC, RxNorm, and SNOMED CT.
- Process Level of HIPAA Security/Privacy, Master Patient Index (MPI), Health Information Exchange (HIE), and so on.
- Tablet PCs, Application Service Provider (ASP) models, Personal Digital Assistants (PDAs), Bar Coding, and other devices at the device level.

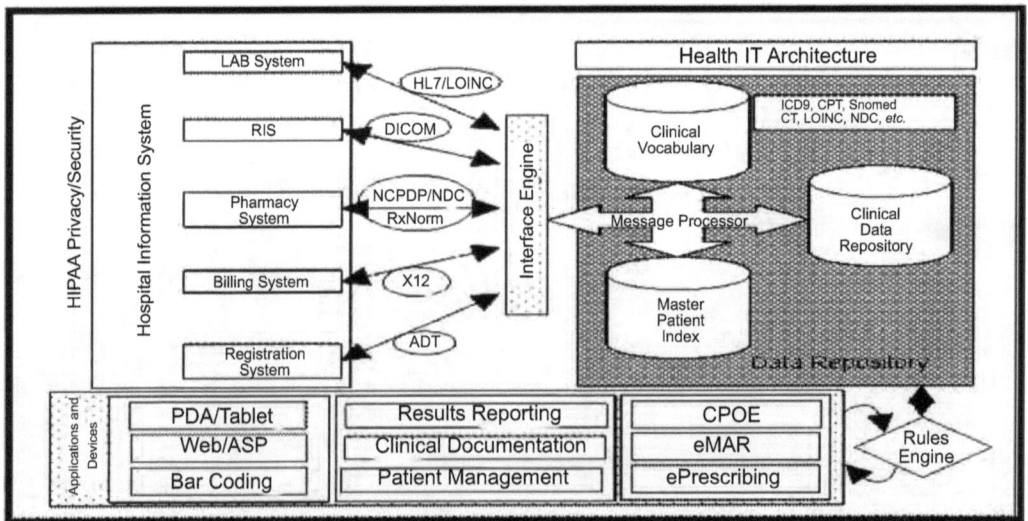

Fig. (1). Architecture of health IT system.

CODING NAMED STANDARDS

The current U.S.-named health language standards established through the regulatory process under HIPAA or promoting interoperability are listed in this section. The main goal of each identified health terminology standard is described in this short overview [12 - 14] (Table **1**).

Table 1. The list of terminology and code set standards for U.S. regulations and requirements for electronic health data exchange [13].

Sr. No.	Terminology Named	Purpose
1.	CDT	Dental services must be recorded in patients' medical records and reported as procedures on claims made to benefit plans in accordance with the HIPAA requirement for Level II "D" codes.
2.	CPT	Billing for medical services and other professional operations, as well as for inpatient services and outpatient treatments.
3.	HCPCS-Level II	Billing for products, supplies, and procedures not included in CPT.
4.	ICD-10	Reporting on mortality
5.	ICD-10-CM	The basis for diagnosis-related groups and other grouper-based payment models in conjunction with ICD-10-PCS is morbidity in outpatient, inpatient, and other care settings.
6.	ICD-10-PCS	Hospital reporting of inpatient operations.
7.	LOINC	Observations, measurements, and records from clinical and laboratory settings are exchanged electronically.
8.	NDC	An FDA-approved product identifier for packaged prescription pharmaceutical goods.
9.	RxNorm	Sharing information about medications
10.	SNOMED CT	Include information on medical equipment, allergies, history, procedures, problem lists, and some clinical findings (such as smoking status) in your records.

BELOW IS A SIMPLE DESCRIPTION OF EACH TERMINOLOGY.

Current Dental Terminology (CDT)

The Code on Dental Procedures and Nomenclature, often known as CDT Current Dental Terminology, is owned and maintained by the American Dental Association. The dissemination of CDT codes is possible both domestically and internationally in order to promote precise, uniform, and consistent reporting of dental treatment. They are used in both paperless and paper-based patient records as well as electronic and paperless dental claims. The code set is identified as

CDT and is a part of Level II of the Healthcare Common Procedure Coding System, as was already mentioned. Every claim made on an electronic dental claim utilizing a HIPAA standard must utilize dental procedure codes from the CDT Code version that was in effect when the service was delivered since CDT is a designated HIPAA standard [23].

CPT, or Current Procedural Terminology

The CPT Current Procedural Terminology (CPT) standard was developed, is copyright protected, and is maintained by the American Medical Association (AMA). The language is most prevalent in the US, while it is also used in other countries. The outpatient services and treatments that hospitals, healthcare organizations, and physicians perform are reported using the code set. CPT must possess a license. The terminology refers to the HIPAA Standard as HCPCS Level I [17].

Healthcare Common Procedure Coding System (HCPCS)

Both public and private healthcare programs utilize the HCPCS Healthcare Common Procedure Coding System (HCPCS). Level I and Level II make up the two main HCPCS subsystems. HCPCS Level I incorporates Current Procedural Terminology (CPT), which is maintained by the American Medical Association (AMA) and is generally used to denote medical services and procedures offered by doctors and other healthcare providers, as well as by hospitals in outpatient settings. HCPCS Level II is a standardized code system maintained by CMS that is primarily used to identify goods, services, and supplies that are not covered by CPT codes. The American Dental Association (ADA) owns the rights to the Current Dental Terminology (CDT) codes, which are a different class of national codes and are classified as HCPCS Level II codes. HCPCS Level III are local codes for the HCPCS system that have been produced by state Medicaid programmes, Medicare contractors, and private insurers for usage in particular programmes and areas. By October 2002, Level III local codes would no longer be in use, as per the regulations, and Level I and Level II code sets would then be available. Section 532(a) of the Biometric Information Privacy Act (BIPA) allowed local codes to be used until December 31, 2003, postponing their abolition [18].

ICD-10

The World Health Organization is the owner of the ICD-10 or 10th Revision of the International Statistical Classification of Diseases and Related Health Problems (WHO). When collecting, classifying, processing, and presenting mortality statistics, it is utilized to encourage cross-national comparability.

Tracking epidemiological trends is done using it. ICD-9 was substituted by the 10th Edition. It has been used in the United States since 1999 and is a recognized standard for coding the cause of death from death certificates (statutory basis) [19].

ICD-10-CM

The ICD-10 modification by the American government has been given the WHO's approval. The usage of the 10th revision of the International Statistical Classification of Diseases and Related Health Problems (ICD-10) in the United States is handled by the National Center for Health Statistics (NCHS), a U.S. Federal agency. For the purpose of modifying the classification clinically, NCHS has done so (ICD-10-CM). ICD-10-CM and the ICD-10-CM Official Guidelines for Coding and Reporting are used to code diagnoses in clinical and outpatient settings, and it is a HIPAA standard for coding morbidity in outpatient, inpatient, and other care settings [19].

ICD-10-PCS

The Centers for Medicare and Medicaid Services (CMS) oversees the ICD-10 PCS, or International Classification of Diseases, 10th Revision, and Procedure Coding System (CMS). ICD-10-PCS updates and alterations are under the direction of CMS, a U.S. government organization [19].

LOINC

For reporting procedures carried out in hospital inpatient healthcare settings, hospitals use a coding system. It is a recognized HIPAA standard for hospital reporting of inpatient operations, together with the ICD-10-PCS official guidelines for coding and reporting.

The Regenstrief Institute owns, operates, and has a license for the LOINC system, which is principally supported by the NLM. A widely accepted international standard for cataloging health measures, observations, and records is called LOINC. The language is a designated standard for meaningful use in the United States. The final rule of Meaningful Use Stage 2 states that LOINC should be used for the electronic sharing of laboratory test results, clinical observations including vital signs, social, psychological, and behavioral data, and documentations like care/referral summaries [21].

National Drug Codes (NDC)

For packaged prescription drug items that have been approved for human consumption; NDC National Drug Codes (NDC) serve as a uniform product identifier. A commercial package's size, labeler, and product are indicated by the NDC code. FDA assigns the third segment, while the product labeler or manufacturer allocates the first two parts of the code. The National Drug Code Directory is a publication of the FDA. All OTC and prescription drug containers and inserts in the United States contain the NDC number. It differs based on the payer and is used for billing and reimbursement. The NDC Directory listing and information are also used in the implementation and enforcement of the act. It is a recognized HIPAA requirement [24].

RxNorm

RxNorm is a tool for promoting semantic interoperability between drug terminologies and pharmacy knowledge base systems. It provides a normalized naming scheme for both generic and branded medications. The NLM is the one who owns it and keeps it up. Despite being accessible internationally, the nomenclature represents the American drug market. American Meaningful Use has a named standard known as RxNorm. In the sharing of pharmaceutical data and medication allergies, RxNorm is specified in the final Meaningful Use Stage 2 regulation [22].

SNOMED-CT

commonly known as a common reference terminology, is a comprehensive, multilingual, regulated clinical reference vocabulary that is used to capture clinical data. It includes a broad spectrum of ailments, clinical revelations, etiologies, techniques, living objects, and results. In order to capture, share, and aggregate health data across specialties and sites of care, SNOMED-CT offers a common language. It has ideas related to clinical expertise to allow for clear and accurate data recording. For therapeutic applications, it can be used to index, store, and retrieve patient information. SNOMED-CT is intended for use in electronic health record systems, not paper-based ones. Clinical data entered into the electronic health record will be encoded behind the scenes by SNOMED-CT codes, which will be integrated into EHR systems. Attempting to manually assign SNOMED-CT codes is not realistic [20].

Additional information on health terminology standards includes the standards' coverage, overlaps, development and upkeep, collaborations, and dissemination. Comparative details of health terminologies are shown one by one in the following tables.

Comparative Analysis of Health Terminologies

CDT (Current Dental Terminology)

Table **2** gives a brief description of Terminology with description.

Table 2. Terminologies [20].

Name	CDT (Current Dental Terminology) https://www.ada.org/en/publications/cdt
Purpose	The goal of the CDT Code is to establish accuracy in documenting dental treatment by ensuring uniformity, consistency, and specificity.
Usage	Distribution of CDT codes is limited to the US and other countries.
Ownership	American Dental Association (ADA)
Area of Use	Oral Health and Dental Services
Key Applications	Documenting Dental Treatment
Coverage	Dental procedures and services.
~Codes	Range from 00100 to 99499
Release	Annual update, effective January 1 of each year
Example	D7250 code for removal of residual roots requiring bone removal.

CPT (Current Procedural Terminology) [25].

Table **3** discuss about the procedural terminology.

Table 3. Current Procedural Terminology [25].

Name	CPT95 (Current Procedural Terminology) https://www.ama-assn.org/practice-management/cpt-current-procedural-terminology
Purpose	Physicians and other healthcare professionals utilize a code set to report freestanding and hospital outpatient medical operations and services.
Usage	Mostly American, but also utilized (but not referred to as a standard) in other nations.
Ownership	American Medical Association (AMA) developed, copyrighted, and maintains the code set.
Area of Use	Medical procedures and services.
Key Applications	Treatment Tracking Billing
Coverage	Services and processes in medicine.
~Codes	10000
Release	Issued once a year, but some category I codes (such as vaccination goods and molecular pathology) are "early released" to take effect six months before the official release date. Other category codes may also be "early released" to catch reporting for evaluation.

(Table 3) cont.....

Example	The code 90387, for example, is described as "Individual Psychotherapy. 60 minutes."

HCPCS (Healthcare Common Procedure Coding System) [26]

Table **4** discuss about the HCPC System terminology and description.

Table 4. HCPC system [26].

Name	HCPCS-Level II96 (Healthcare Common Procedure Coding System) https://www.cms.gov/Medicare/Coding/MedHCPCSGenInfo/index.html
Purpose	Codes that are used to identify goods, services, and things when they are used outside of a doctor's office, like ambulances and durable medical equipment and supplies.
Usage	In the United States, non-physician-based services and commodities are identified and billed using the standardized coding system known as HCPC-Level II.
Ownership	In accordance with the HIPAA regulations, the Secretary of HHS granted CMS permission to preserve and disseminate HCPCS Level II Codes in October 2003.
Area of Use	When utilized outside of a doctor's office, ambulance services, sturdy medical equipment, prosthetics, orthotics, and supplies.
Key Applications	Billing Medicare and Medicaid
Coverage	Products, supplies, devices and services are not covered by CPT
~Codes	7404
Release	Quarterly with an annualized compilation.
Example	C1729 Catheter, drainage

ICD-10 (International Statistical Classification of Diseases -10)

Table **5** discuss about the statistical clarification.

Table 5. ICD of diseases [26].

Name	ICD-10 (International Statistical Classification of Diseases and Related Health Problems 10th Revision) http://www.who.int/classifications/icd/en/
Purpose	To encourage inter-national comparability in the gathering, grouping, processing, and display of mortality information. It is employed to monitor epidemiological trends and support choices on medical reimbursement.
Usage	Worldwide with regional/national variations (for U.S. see ICD-10-CM below) available in translations in 42 different languages.
Ownership	World Health Organization (WHO)
Area of Use	Diseases and Diagnoses

(Table 5) cont.....

Key Applications	Statistics Billing
Coverage	Covers the two key components of the health system that are associated with mortality and morbidity.
~Codes	68000
Release	Every three years for significant changes, and once a year for smaller ones, the tabular list is updated.
Example	Example Right ankle sprain, initial encounter: S93.401A, S93.401a, s93.401A, s93.401a)

ICD-10-CM (International Statistical Classification of Diseases - Clinical Modifications) [27]

Table **6** ICD-10-CM (International Statistical Classification of Diseases - Clinical Modifications) highlighting key aspects:

Table 6. Clinical modification [27].

Name	ICD-10-CM97 (International Statistical Classification of Diseases - Clinical Modifications) https://www.cdc.gov/nchs/icd/icd10cm.htm
Purpose	Clinical Modification of ICD-10 unique to the US. Used in clinical and outpatient settings for diagnosis coding.
Usage	U.S. coding system: utilized internationally (Spain, Belgium).
Ownership	World Health Organization (WHO)
Area of Use	Diseases and Diagnoses
Key Applications	Statistics Billing
Coverage	Covers the two key components of the health system that are associated with mortality and morbidity.
~Codes	69100
Release	Annual updates effective from October 1 of each year.
Example	**ICD-10-CM Code Examples** • I25.110, Arteriosclerotic heart disease of the native coronary artery with unstable angina pectoris. • K50.013, Crohn's disease of small intestine with fistula. • K71.51, toxic liver disease with chronic active hepatitis with ascites.

ICD-10-PCS (International Statistical Classification of Diseases - Clinical Modifications) [28] (Table 7)

Table 7. ICD-10-PCS [28].

Name	ICD-10-PCS99 ICD-10 (Procedure Coding System) https://www.cms.gov/Medicare/Coding/ICD10/2019-ICD-10-PCS.html
Purpose	U.S. system of codes for hospital reporting of procedures carried out in inpatient hospital settings.
Usage	U.S. coding system: utilized internationally (Portugal, Spain, and Taiwan).
Ownership	Medicare and Medicaid Services (CMS). The U.S. government organization in charge of regulating all alterations to the ICD-10-PCS is called CMS.
Area of Use	System, which is used for coding procedures and services provided in the inpatient setting of hospitals in the United States.
Key Applications	Procedure Coding System
Coverage	In-patient procedures.
~Codes	87,000
Release	Annual addenda October 1.
Example	0HQEXZZ Medical and Surgical section (0), body system Skin and Breast (H), root operation Repair (Q), body part Skin, Left Lower Arm (E), External approach (X) No Device (Z) and No Qualifier (Z).

LOINC (Logical Observation Identifiers, Names and Codes) [21]

Table **8** discuss about LOINC (Logical Observation Identifiers, Names and Codes) a standardized coding system.

Table 8. LOINC [21].

Name	LOINC (Logical Observation Identifiers Names and Codes) https://loinc.org/
Purpose	A common language for describing health measurements, findings, and records.
Usage	Available in the following translations: Chinese, Dutch, Estonia, Belgian French, Canadian French, French, Swiss French, Greek, Italian, Swiss Italian, Korean, Brazilian Portuguese, Russian, Argentinian Spanish, Mexican Spanish, Spanish, Spanish, Turkish, as well as Austrian German, German, Swiss German, and Greek.
Ownership	The National Library of Medicine provides the majority of the funding for LOINC, which is owned and administered by Regenstrief Institute.
Area of Use	Laboratory orders and results
Key Applications	Transmitting Laboratory and test observations

(Table 8) cont.....

Coverage	Everything about a patient that may be evaluated, measured, or seen falls under the general scope. All findings communicated by clinical laboratory.
~Codes	55,000
Release	Regenstrief creates LOINC data releases. Every year, in June and December, the LOINC nomenclature is released. It is freely accessible through Regenstrief.
Example	LOINC Code System Short Name Long Name Type 29463-7 Patient Weight Body weight Clinical 11450-4

NDC (National Drug Codes) [24]

Table **9** depicts standardize the identification of drugs in electronic health records.

Table 9. NDC [24].

Name	NDC (National Drug Codes) https://www.fda.gov/drugs/informationondrugs/ucm142438.html
Purpose	A generic product number for prescription drugs that is safe to consume. Labeler, product, and commercial package size are indicated by the NDC code.
Usage	Depending on the payer, a different NDC code is used for billing and reimbursement, and the NDC Directory listing and information are utilized to execute and enforce the Act.
Ownership	FDA produces the National Drug Code Directory.
Area of Use	Pharmacy Products.
Key Applications	Drugs reimbursement reporting drugs and biological products.
Coverage	In scope – human prescription drug, OTC, or insulin product.
~Codes	55,000
Release	NDC Directory is updated daily. The downloadable data file is offered in SAS, Stata, and CSV formats.
Example	For example, the NDC for a 100-count bottle of Prozac 20 mg is 0777-3105-02.Labeler, Product Code and Package Code

RxNorm [22]

Table **10** shows about standardized nomenclature for clinical drugs.

Table 10. RxNorms [22].

Name	RxNorm https://www.nlm.nih.gov/research/umls/rxnorm/
Purpose	RxNorm is a normalized naming system for both generic and branded medications as well as a tool to enable semantic interoperability between drug terminologies and pharmacy knowledge base systems.

(Table 10) cont.....

Usage	U.S. - Represents the U.S. medication market that is accessible worldwide.
Ownership	The National Library of Medicine is the owner and operator of RxNorm.
Area of Use	Clinical drugs and drug delivery devices
Key Applications	Recording and Processing drug information
Coverage	RxNorm lists the names of several over-the-counter medicines and prescription medications that are sold in the United States. RxNorm includes branded and generic medications:
~Codes	230643
Release	Newly approved medicine information from the FDA Structured Product Labeling source vocabulary is released weekly (every Wednesday) by RxNorm.
Example	The RxNorm code for ciprofloxacin 500 mg 24-hour extended-release tablet (the generic name for Cipro XR 500 mg) is RX10359383

SNOMED CT (Systematized Nomenclature of Medicine) [20]

Table **11** discuss about comprehensive clinical terminology that provides standardized codes for diseases, clinical findings, procedures, and other healthcare concepts.

Table 11. SNOMED [20].

Name	CT SNOMED was the first meaning of the acronym SNOMED, which was lost in 2002 when SNOMED and CTV3 were united. https://www.snomed.org/
Purpose	Clinical jargon enables the automation of analytical and reasoning processes for processing EHR data.
Usage	Global
Ownership	Organization for the Development of International Health Terminology Standards, doing business as SNOMED International (SI). An international, nonprofit organization is SNOMED International. The representative for the United States is the National Library of Medicine on behalf of the Department of Health and Human Services.
Area of Use	Clinical Terminology
Key Applications	Recording, aggregation and sharing clinical data
Coverage	• Problems are among the clinical findings. • Procedures, which are broadly described as all health-related actions include taking a patient's history, performing a physical exam, administering tests, using imaging technology, performing surgery, providing training and education on specific diseases, counseling, and so on.
~Codes	352,567
Release	The International Release of SNOMED CT is released twice a year – January 31 and July 31.

(Table 11) cont.....

Example	• **Body structure** 123037004. • **Clinical findings** 404684003. • **Context-dependent categories** 243796009

The Graphical Representation of Medical Terminology ~code set and for year of establishment shown in Figs. (**2** and **3**).

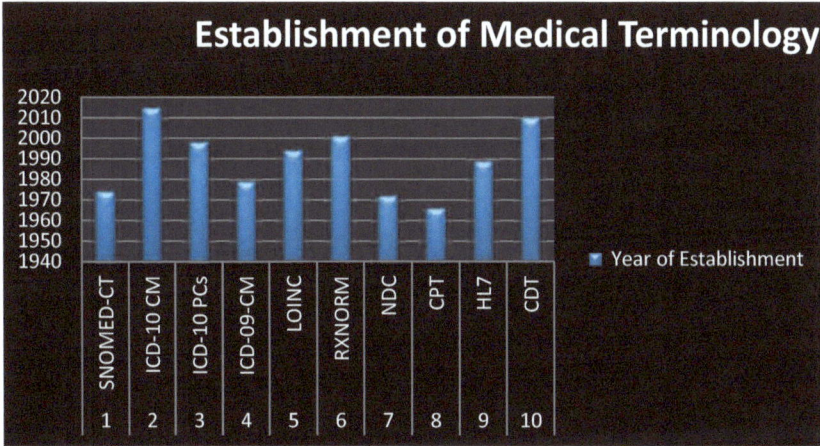

Fig. (2). Medical terminology with year.

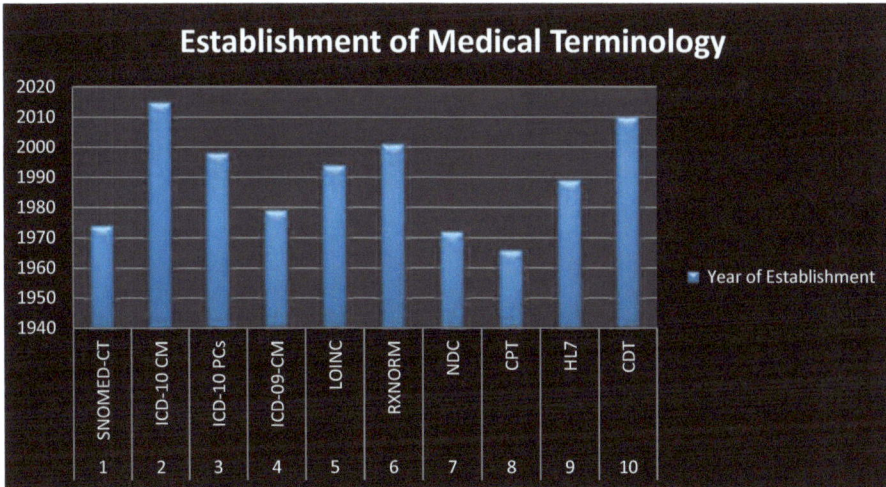

Fig. (3). Medical terminology with ~code set.

DISCUSSION

The goal of this study article is to investigate the main medical jargon that practitioners in the healthcare sector such as physicians, nurses, chemists, and medical assistants frequently utilize. The importance of medical language may be attributed to three key factors: First, it enables straightforward communication between doctors. Second, it gives professionals a standard vocabulary to use when describing illnesses and diseases in patients. Thirdly, it enables medical professionals to oversee and carry out patient care.

Medical terminology refers to the words and phrases used above to describe various biological parts and their activities. It is frequently used to describe symptoms, diseases, and medical diagnoses in professional situations.

The foundations and their use are as follows:

- **Who Uses It:** It is used by physicians, registered nurses, physician assistants, and other medical specialists.
- **What:** It is possible to discuss signs and symptoms, recognized diagnoses, tested treatment modalities, and dose recommendations.
- **When:** It is used by medical experts to identify and treat patients. It is also used by professionals in medical billing and coding operations.
- **Uses:** It is employed in healthcare management organizations, private offices, operating rooms, and hospitals.

Stakeholders and governments from around the world have recently stated that all healthcare companies must use electronic health records (EHRs). Their goals in doing this are to boost patient safety, decrease medical errors, reduce expenses, and improve efficiency [29, 30]. It is impossible to accomplish these goals without the medical knowledge on which the EHR is based. Medical knowledge must be represented using patient data from a variety of sources, including problem lists, progress notes, actions, medication lists, laboratory and complementary test results, social and environmental determinants of health, environmental data, patient decisions regarding their health and medical treatments, and genomics and proteomics data. Natural language is incredibly detailed, yet it also lacks precise meanings, employs jargon and acronyms, depends heavily on context, and is confusing. In order to capture structured data items in databases, many current EHRs use template-based methods. The expressiveness and flexibility that clinicians are used to when entering data are not possible with structured data, and because contextual information is lost, it can be challenging to evaluate and reconstruct the meaning from structured data [32, 44]. From the standpoint of the healthcare professional, data entry is a barrier to EHR uptake and effective utilization. In free writing using everyday language, healthcare professionals like to document observations, procedures, and results [31]. Using a narrative framework, they may swiftly and effectively convey complex ideas.

Therefore, it is necessary to define ambiguities and standardize vocabulary. An EHR must record clinical data in a structured, ideally coded way to achieve this. Reduction is a method used to codify ideas [33]. Codes are often either alphabetic or numeric. Real-world facts must be recorded using a standard codification system (SCS) in order to be controlled in a database.

Evans *et al.* claimed that a "common, uniform, and comprehensive approach to the display of medical information" was required by the medical community [34]. In one-to-one coding, one term should only be present for a certain item. Per phrase, there should only be one object.

Ambiguity is meant to be prevented by adopting homonymy [33]. This must be done in the field of health utilizing international standards, such as terminology (SNOMED-CT) or classifications (ICD 10, CPC, *etc.*) [35, 36]. Following the publication of Cimino's Desiderata [37], the distinction between terminology systems like SNOMED-CT and classification systems like ICD-10-CM/PCS became more obvious. Both coding schemes provide the necessary data structure to enable clinical and administrative processes in the healthcare sector. Systems for clinical classification and nomenclature were initially developed to satisfy the demands of different users and accomplish a variety of objectives [38]. The ICD-10 classification system is a product for general reporting needs, such as, for example, public health surveillance. Contrarily, SNOMED-CT was developed as common data architecture for medical applications [39].

The International Standards Organization (ISO) states that terminologies should be formal aggregations of concepts that are independent of language, with each concept being represented by a preferred term and suitable synonyms, as well as by explicitly shown connections between the concepts [40, 41]. Standard terminologies should be used in an EHR to assist decision support systems [36, 42, 43], data sharing between health information systems, epidemiological analysis, research to support health services research, administrative job management, and other purposes.

CONCLUSION

The objective of the research is to find out the best healthcare terminology, out of RxNORM, LOINC, NDC, and SNOMED Clinical Terms (SNOMED- CT). They also include ICD-10-CM, ICD-10-PCS, and ICD-10, as well as CDT, CPT, HCPCS-Level II to achieve interoperability, security and privacy needed in the healthcare domain, with the appropriate use. As per the background and comparison study, SNOMED- CT seems to be the best healthcare terminology compared to others. The comparison study covered all the attributes in the form of standards' coverage, overlaps, development and upkeep, collaborations, and dissemination . We analyzed all the above mentioned terminologies and found SNOMED- CT is the better choice among others. SNOMED CT enabled clinical health records benefit populations by facilitating early identification of emerging health issues, monitoring of population health, and agile response to changing clinical practices, thereby enabling accurate access to relevant information, and reducing costly duplications and errors.

Although the ideas of clinical terminology and classification are simple to grasp, scaling them up for an IT department is challenging. Healthcare businesses need to be prepared to give up spreadsheets and devote time and money to a specialized terminology solution in order to prevent billing errors and clinical mishaps.

REFERENCES

[1]   Tiwari, "NLP in Healthcare", *Artificial Intelligence Magazine,* 2020.

[2]   World Health Organization, "Classifications" [Internet]. Geneva, Switzerland: World Health Organization; updated Feb. 23, 2018; cited May 2018.

[3]   World Health Organization, "Classifications" [Internet]. Geneva, Switzerland: World Health Organization; updated Feb. 23, 2018; cited May 2018.

[4]   S. B. Kanaan, "The national committee on vital and health statistics 1949-99: A history", hyattsville, MD: Centers for disease control and prevention: national center for health statistics (U.S.), Sep. 2001, p. 33.

[5]   Institute of Medicine, *The Computer-Based Patient Record: An Essential Technology for Health Care.* The National Academies Press: Washington, DC, 1991.

[6]   Centers for Medicare and Medicaid Services, "New CMS coding changes will help beneficiaries: Improvements will speed use of technology", *CMS Medicare News,* 2004.

[7]   "Social Security Amendments of 1983," H.R. 1900/P.L.98-21, Apr. 20, 1983.

[8]   "Case-mix measurement and assessing quality of hospital care" [Internet]. *Health Care Financing Review, Suppl.,* 1987 [cited Mar. 2018], pp. 39–48.

[9]   President's Advisory Commission on Consumer Protection and Quality in the Health Care Industry, Final Report [Internet]. Rockville, MD: Agency for Healthcare Research and Quality; Mar. 12, 1998 [cited Aug. 2018].

[10]   National Quality Forum, "What We Do" [Internet]. Washington, DC: National Quality Forum; updated 2018; cited Aug. 2018.

[11]   National Committee on Vital and Health Statistics, *Report to the Secretary of the U.S. Department of Health and Human Services on Uniform Data Standards for Patient Medical Record Information.* National Committee on Vital and Health Statistics: Washington, DC, 2000. Internet

[12]   Centers for Medicare and Medicaid Services, "CMS Value-Based Programs" [Internet]. Baltimore, MD: U.S. Centers for Medicare and Medicaid Services; updated Jun. 2018; cited Jul. 2018.

[13]   Centers for Medicare and Medicaid Services, "Promoting Interoperability (PI)" [Internet]. Baltimore, MD: U.S. Centers for Medicare and Medicaid Services; updated May 31, 2018; cited Jul. 2018,

[14] The Office of the National Coordinator for Health Information Technology, "Draft Trusted Exchange Framework" [Internet]. Washington, DC: The Office of the National Coordinator for Health Information Technology; updated Jan. 5, 2018; cited Jul. 2018.

[15] The Office of the National Coordinator for Health Information Technology, Certification Guidance for EHR Technology Developers Serving Health Care Providers Ineligible for Medicare and Medicaid EHR Incentive Payments [Internet]. Washington, DC: The Office of the National Coordinator for Health Information Technology (U.S.), Sep. 9, 2013 [cited Jul. 2018].

[16] E. S. Anthony and M. L. Lipinski, "2015 edition final rule: overview of the 2015 edition health it certification criteria & onc health it certification program provisions" [Internet]. Washington, DC: The Office of the National Coordinator for Health Information Technology; Oct. 28, 2015 [cited May 2018].

[17] The American Medical Association, "CPT (Current Procedural Terminology)" [Internet]. Chicago, IL: The American Medical Association; cited May 2018.

[18] Centers for Medicare and Medicaid Services, Healthcare Common Procedure Coding System (HCPCS) Level II Coding Procedures [Internet]. Baltimore, MD: Centers for Medicare and Medicaid Services (U.S.), 2015 [cited Jun. 2018], p. 12.

[19] Centers for Medicare and Medicaid Services and National Center for Health Statistics, ICD-10-CM Official Guidelines for Coding and Reporting FY 2018 [Internet]. Atlanta, GA: Centers for Disease Control and Prevention (U.S.), 2018 [cited Mar. 2018], p. 117.

[20] D. Kalra et al., "Assessing SNOMED CT for Large Scale eHealth Deployments in the EU: ASSESS CT Recommendations" [Internet]. Bonn: ASSESS CT (DE), Dec. 2016 [cited Jul. 2018], p. 23.

[21] W.S. Campbell, D. Karlsson, D.J. Vreeman, A.J. Lazenby, G.A. Talmon, and J.R. Campbell, "A computable pathology report for precision medicine: extending an observables ontology unifying SNOMED CT and LOINC", *J. Am. Med. Inform. Assoc.,* vol. 25, no. 3, pp. 259-266, 2018. [http://dx.doi.org/10.1093/jamia/ocx097] [PMID: 29024958]

[22] RxNorm (Liu, Ma, Moore, Ganesan, & Nelson, 2005), "A system for normalizing drug names and supporting interoperability between health systems."

[23] American Dental Association, "Current Dental Terminology (CDT)" [Internet]. Archived from the original (web) on Nov. 19, 2010; retrieved Oct. 17, 2010.

[24] Woolstenhulme, "NDC Codes and Drug Classification Systems," Apr. 20, 2020.

[25] S. Panchbhai, and V.M. Pathak, "A systematic review of natural language processing in healthcare", *J. Algebr. Stat.,* vol. 13, no. 1, pp. 682-707, 2022.

[26] Centers for Medicare and Medicaid Services, *Healthcare Common Procedure Coding System (HCPCS) Level II Coding Procedures.* Centers for Medicare and Medicaid Services (U.S.): Baltimore, MD, 2015.

[27] Centers for Medicare and Medicaid Services and National Center for Health Statistics, *ICD-10-CM Official Guidelines for Coding and Reporting FY 2018.* Centers for Disease Control and Prevention (U.S.): Atlanta, GA, 2018.

[28] P. Brooks and R. Butler, "ICD-10-PCS," presented at: National Committee on Vital and Health Statistics Full Committee Hearing [Internet]; Jun. 21–22, 2017 [cited Mar. 2018].

[29] D.W. Bates, R.S. Evans, H. Murff, P.D. Stetson, L. Pizziferri, and G. Hripcsak, "Detecting adverse events using information technology", *J. Am. Med. Inform. Assoc.,* vol. 10, no. 2, pp. 115-128, 2003. [http://dx.doi.org/10.1197/jamia.M1074] [PMID: 12595401]

[30] W. Bates, D.L. Boyle, and J.M. Teich, "Impact of computerized physician order entry on physician time", *Proceedings of the Annual Symposium on Computer Applications in Medical Care,* pp. 996-1000, 1994.

[31] A.L. Rector, "Clinical terminology: why is it so hard?", *Methods Inf. Med.,* vol. 38, no. 04/05, pp. 239-252, 1999.
[http://dx.doi.org/10.1055/s-0038-1634418] [PMID: 10805008]

[32] J.S. Shapiro, "Document ontology: Supporting narrative documents in electronic health records", *AMIA Annual Symposium Proceedings,* pp. 684-688, 2006.

[33] K. J. Hannah and M. J. Ball, Health Informatics. (Formerly Computers in Health Care).

[34] A. Evans et al., "Toward a medical-concept representation language," 1994.
[http://dx.doi.org/10.1136/jamia.1994.95236153]

[35] K. Giannangelo, Healthcare Code Sets, Clinical Terminologies, and Classification Systems, 2nd ed., Chicago, IL: American Health Information Management Association (AHIMA), 2015.

[36] C.G. Chute, "Clinical classification and terminology: some history and current observations", *J. Am. Med. Inform. Assoc.,* vol. 7, no. 3, pp. 298-303, 2000.
[http://dx.doi.org/10.1136/jamia.2000.0070298] [PMID: 10833167]

[37] J.J. Cimino, "Desiderata for controlled medical vocabularies in the twenty-first century", *Methods Inf. Med.,* vol. 37, no. 04/05, pp. 394-403, 1998.
[http://dx.doi.org/10.1055/s-0038-1634558] [PMID: 9865037]

[38] Z.M. Alakrawi, "Clinical terminology and clinical classification systems: A critique using AHIMA's data quality management model", *Perspect. Health Inf. Manag.,* 2016. [Internet].

[39] Perspectives in Health Information Management, "Fall 2014 introduction" [Internet], vol. 11, 2014.

[40] International Standards Organization (ISO), ISO/TS 17117:2002(E): Health Informatics-Controlled Health Terminology—Structure and High-Level Indicators. Technical Committee ISO/TC 215, Health Informatics, 2002.

[41] International Standards Organization (ISO), TC 215—Health Informatics, 2007.

[42] J.J. Cimino, "Terminology tools: state of the art and practical lessons", *Methods Inf. Med.,* vol. 40, no. 4, pp. 298-306, 2001.
[http://dx.doi.org/10.1055/s-0038-1634425] [PMID: 11552342]

[43] E.T. Wong, T.A. Pryor, S.M. Huff, P.J. Haug, and H.R. Warner, "Interfacing a stand-alone diagnostic expert system with a hospital information system", *Comput. Biomed. Res.,* vol. 27, no. 2, pp. 116-129, 1994.
[http://dx.doi.org/10.1006/cbmr.1994.1012] [PMID: 8033537]

[44] B. S. Panchbhai and Dr. Varsha M. Pathak, "A systematic review of natural language processing in healthcare," Journal of Algebraic Statistics, vol. 13, no. 1, pp. 682–707, May 2022, ISSN: 1309-3452.

SUBJECT INDEX

www.ingramcontent.com/pod-product-compliance
Lightning Source LLC
Chambersburg PA
CBHW050820220326
41598CB00006B/275